# Unmanned System Technologies

Springer's Unmanned Systems Technologies (UST) book series publishes the latest developments in unmanned vehicles and platforms in a timely manner, with the highest of quality, and written and edited by leaders in the field. The aim is to provide an effective platform to global researchers in the field to exchange their research findings and ideas. The series covers all the main branches of unmanned systems and technologies, both theoretical and applied, including but not limited to:

- Unmanned aerial vehicles, unmanned ground vehicles and unmanned ships, and all unmanned systems related research in:
- Robotics Design
- Artificial Intelligence
- Guidance, Navigation and Control
- Signal Processing
- Circuit and Systems
- Mechatronics
- Big Data
- Intelligent Computing and Communication
- Advanced Materials and Engineering

The publication types of the series are monographs, professional books, graduate textbooks, and edited volumes.

To submit a proposal or request further information, please contact:
Dr. Mengchu Huang, Senior Editor, Applied Sciences
Email: mengchu.huang@springer.com
Tel.: +86-21-2422 5094

More information about this series at http://www.springer.com/series/15608

Xin Bi

# Environmental Perception Technology for Unmanned Systems

華中科技大學出版社
Huazhong University of Science and Technology Press

Springer

Xin Bi
Tongji University
Shanghai, China

ISSN 2523-3734 ISSN 2523-3742 (electronic)
Unmanned System Technologies
ISBN 978-981-15-8095-6 ISBN 978-981-15-8093-2 (eBook)
https://doi.org/10.1007/978-981-15-8093-2

Jointly published with Huazhong University of Science and Technology Press
The print edition is not for sale in China (Mainland). Customers from China (Mainland) please order the print book from: Huazhong University of Science and Technology Press.



This Springer imprint is published by the registered company Springer Nature Singapore Pte Ltd.
The registered company address is: 152 Beach Road, #21-01/04 Gateway East, Singapore 189721, Singapore

# Preface

With the rapid socio-economic development and the continuous advancement of the artificial intelligence industry, the autonomous unmanned systems have become an integral part of the human life and the production line. For instance, the unmanned aerial vehicles are deployed to perform electrical inspection, movies and television filming, map surveying, air traffic control, etc; the unmanned vehicles are used for logistics transportation, public transportation guidance, etc; and the unmanned surface vehicles are used in marine environment monitoring, seabed surveying, sea area monitoring and rescue, etc. Each application of the autonomous unmanned systems has brought about great convenience and enhanced safety in our lives. With the system safety technology as the precondition, the operation of the unmanned aerial vehicle, the unmanned vehicle and the unmanned surface vehicle are controlled through a variety of the advanced auxiliary systems. The key factor for achieving its safety lies in the use of the environment-sensing system in the unmanned aerial vehicle, the unmanned vehicle and the unmanned surface vehicle which obtain the environmental information around the autonomous unmanned system. Hence, the environment awareness technology has become an important research field in the autonomous unmanned system and is a research focus for many researchers in the various research organizations.

In the process of sensing the environment, the autonomous unmanned system continuously senses the environment through the sensors that are installed in the system. It detects stationary and moving objects in the surrounding environment to further track and recognize the target. Through the environmental awareness and understanding algorithms, the target's movement and threat level is determined. The system then automatically plans and executes the corresponding actions. The research purpose of the environmental sensing is to understand the unmanned system's ability to collect and recognize the various types of the relevant spatial information; understand the surroundings situation; and subsequently guide the decision-making and action to be taken. Pertaining to the unmanned vehicles, the objective of the environmental perception is based on the driving performance and consensus rules that can identify objects with potential safety hazard in the surroundings of the driving path in real time, reliably and accurately. It is used to plan out the driving path that can ensure the safe and rapid arrival at the destination and take the necessary actions for itself to avoid traffic accidents. In the perspective of the unmanned surface vehicle, similar to the unmanned vehicle, the objective of the environmental perception is to obtain the surface navigation situation in real time. The central processing unit is based on the sensing, planning and decision results to implement the corresponding actions.

This book strives to discuss the theoretical research and practical applications that focus on the sensors and the algorithms that are involved in the environmental perception. The theoretical research and its application in the autonomous unmanned systems are based on the author's research outcome in the millimetre-wave radar, data fusion, etc., that integrates the complex networks and the artificial intelligence concepts and methodologies to build a system of methodologies for sensing the environment. This book provides an overview of the development of the autonomous unmanned systems and the key technologies that are involved. It also depicts a detailed introduction to the millimetre-wave radar technology, the Lidar technology, the machine vision technology, the infrared sensor technology and the ultrasonic sensor technology. A detailed explanation is subsequently provided on the sensor fusion technology, positioning and navigation technology and the autonomous path planning technology. Each chapter is organized into three sections: (1) Basic concepts and principles; (2) Classic and advanced algorithms based on the advancement of the technology in the recent years; and (3) Applications of the basic concepts, principles and algorithms in the autonomous unmanned systems.

The author and the research team had conducted many years of in-depth research in the field of the environmental perception, millimetre-wave radar technology, sensor fusion technology and other relevant technologies of the autonomous unmanned system and had undertaken numerous national- and ministerial-level scientific research topics and projects, which earned them a wealthy accumulation of experience and achievements. Majority of the content comes from these experiences and achievements, which also includes the content from many original research journals. These experiences and achievements provide enriched materials and the basis for the application.

In the booming era of the artificial intelligence, environmental awareness technology will only become more significant. Due to the limited resources and time, this book only comprises of research and exploration in the field of environmental awareness perception and its application in the autonomous unmanned systems. We welcome the readers to provide feedback for us to improve the content of the book. The author will continue to conduct in-depth research into the internal mechanism of the environmental awareness technology to keep up with the rapid advancement of the technology in the autonomous unmanned systems.

Last but not least, my sincere gratitude and thanks to all those who had helped in the process of compiling this book and to the people who had contributed in improving and revising the content to ensure the academic quality of the book.

Shanghai, China
August 2019

Xin Bi

# Contents

# Chapter 1
# Overview of Autonomous Unmanned Systems

## 1.1 Introduction

Since the beginning of the 21st century, research in the intelligent systems had gained tremendous popularity as various technologies such as the micro-electronics technology, the intelligent sensor technology, the advanced intelligent control technology and other information technology, especially with the recent breakthrough in the field of deep learning, neural network and other artificial intelligence, had seen incredible advancement. It is worthwhile to note that the development of the intelligent technology is closely related to the current low-cost and high-speed data computing and storage capabilities. For instance, the network communication technology that had been applied in the military, industrial, medical and other relevant fields, had proven outstanding performance which result in the increased trust, and hence, attention on the intelligent unmanned systems. It can be predicted that the intelligent unmanned systems will become the commanding point of science and technology in the near future and an indicator of a country's scientific and technological international status.

Unmanned systems initially served the military on the battlefields. Needless to say, the military continues its keen attention in the development of the unmanned systems such as the ground unmanned combat systems, the unmanned aerial systems and the marine unmanned systems. Some key reasons of such earnest interest in the intelligent unmanned systems is because of its durability and precision in combat and the ability to overcome the harshness of the combat environment and other aspects which a human combat personnel could never achieve. This was ascertained via the U.S. military's unmanned system's performance on the battlefields of Iraq and Afghanistan. Over the past decade or so, unmanned systems have played an increasingly important role in the U.S. military operations, from the deep underwaters to great heights, and from matchbox size to the size of a Boeing 737. However, regardless of the location and size, the intelligent unmanned systems had



X. Bi, *Environmental Perception Technology for Unmanned Systems*, Unmanned System Technologies, https://doi.org/10.1007/978-981-15-8093-2_1

demonstrated excellent ability and performance to take on more risks than what the humans could afford to, liberating tremendous energy.

Ever since the U.S. Department of Defense adopted the unmanned technology, the unmanned aerial vehicle (UAV) system, or commonly known as the drone, had received the most attention in the aviation space. It is anticipated that the investment in the UAV systems will continue to make up for the majority of the U.S. Defense Department's military budget. In fact, a large number of UAV systems that are capable of carrying out a wide range of missions were already set for deployment. Initially, the key agenda of deploying the drone system was primarily for tactical reconnaissance, but very quickly, this had been extended to cover the Intelligence, Surveillance, and Reconnaissance (ISR) including the surveillance of the warfare condition on the battlefield. UAV systems play an increasingly important role in attack missions as the military is equipped with a variety of weapon systems that are capable of striking important targets.

Since the beginning of the operations in Iraq and Afghanistan, unmanned ground systems have supported a range of operations such as mobility, mobility support and long-term support. Mobility operations concentrate in using speed and firepower to approach and destroy the enemies. Mobility support tasks focus in reducing natural and artificial obstacles and hazards. Long-term support refers to the operational support (utilize, maintain, support and combat service guarantee) that the unmanned ground vehicles could provide. Approximately 8000 unmanned vehicles of various functionalities were deployed in the "Operation Enduring Freedom" and "Operation Iraqi Freedom". As of September 2010, these deployed unmanned ground vehicles had carried out more than 125,000 missions, including target identification and road clearing, positioning and the removal of improvised explosive devices. During the mission to remove the improvised explosive devices, the Army, Navy and Marine Corps Explosive Ordnance Demolition Team used the unmanned ground vehicles to detect and destroy more than 11,000 improvised explosive devices.

Like the drone systems and unmanned ground systems, the unmanned marine systems have the potential to save lives, reduce human risk, provide long-lasting surveillance and reduce operational costs. Since the U.S. military in 1997 put forward the concept of "Network Centric Warfare", the "Undersea Network Centric Warfare" program in the U.S. Navy has been highly valued. The U.S. Navy recognizes that a strong underwater dominance can provide strategic and war advantages for the U.S. defense security, and in order to maintain the U.S. underwater warfare dominance and take full advantage of the new technologies, the U.S. attaches great importance to the development of unmanned combat systems based on underwater tactical networks. The U.S. Department of Defense had released four editions of the "Unmanned Systems Roadmap" in 2007–2013, which sets out a top-level development plan for all types of unmanned systems in the U.S. military [6]. The Defense Advanced Research Projects Agency (DARPA) had invested in the Distributed Agile Submarine Hunting (DASH) project which intended to develop two subsystems in the deep and shallow seas respectively [9]. The deep-sea subsystem consists of the fixed sonar node and dozens of UUVs on the sea floor. The innovation adopts the bottom-up detection method which overcomes the problem of indistinct target information that is caused by the refraction of the sound waves of the sea floor and the sea surface, and effectively reduces the influence of the sea floor terrain on acoustic detection. The shallow sea subsystem utilizes the unmanned system that is equipped with non-acoustic sensors, which monitors shallow-sea submarines from high altitudes to collect non-acoustic features such as the submarine tails.

## 1.2 Development of Unmanned Systems

Since the birth of the first unmanned aerial vehicle in 1917, the development of unmanned systems technology has gone through more than 100 years of history, which can be classified into the exploratory stage, the developing stage and the booming stage.

1. **Exploratory Stage**

1917–1956: Aerial vehicles that were invented before 1935 were unable to return to the starting position until the invention of the Bee King, which allowed the drone to return to the take-off locations. The famous Avenger I was created in 1944 which can carry up to 2000 lb of missiles. In the 1950s, the Soviet Union used small remote-controlled unmanned surface boats to launch suicide attacks against enemy ships. In August 1925, the first endorsed driverless car in human history was officially unveiled. Francis P. Houdina, an electronics engineer from the U.S. Army, sits in a car that uses radio control to manipulate the car in front of him. Radio waves were transmitted by the following car to control the steering wheel, clutch, brakes and other vehicle components of the leading vehicle. Later in 1956, GM officially unveiled the Firebird II concept car which is the world's first concept car that is equipped with a car safety and automatic navigation system (Fig. 1.1).

2. **Developing Stage**

1956–2005: The Vanguard series of drones, which first flew in December 1986 is still in service, taking off with rocket power, weighing 416 lb and traveling at 109 miles per hour. The aircraft can float on the surface of the water, as well as, be recovered by landing on the water surface. In the 1960s, China repaired the

**Bee King** **Avenger I** **Firebird II concept car**

**Fig. 1.1** Key unmanned systems during the exploratory stage

AQM-34N Firebee unmanned reconnaissance aircraft that crashed in China and northern Vietnam, and subsequently produced the WZ-5 Changhong drone which became the starting point of the Chinese drone technology evolution. In 2002, the China Airlines Group which belonged to the Guizhou Aviation Industry (Group) Co., Ltd. had developed the WZ-2000 "Thousand Mile Eye" invisible unmanned reconnaissance aircraft. The 2004 RQ-7B Phantom is the smallest in the drone family, and the system is capable of locating and identifying targets at 125 km away from the Tactical Command Center, allowing commanders to observe, direct, and act more nimbly. In the 1970s and 1980s, due to technical limitations, the development of the unmanned surface boats did not achieve any significant breakthrough. The marine unmanned system was mainly used in military exercises and for artillery firing at offshore targets. Entering the 21st century, with the development of the communication, artificial intelligence and other relevant technologies, many technical bottlenecks which had restricted the development of the unmanned boats were partially resolved. Numerous countries have since increased the research and development in the unmanned boats, and hence, facilitated a period of high-speed development. Shakey, which was developed by the SRI Atificial Intelligence Research Center of Stanford University Institute in the United States, pioneered the autonomous navigation in 1966. In 1977, the Tsukuba Engineering Research Laboratory in Japan developed the first camera-based cruise system. Other than the U.S., in the 1970s and 1980s, Ernst Dickmanns, an aerospace professor from Bundeswehr University in Munich, Germany, had pioneered a series of research projects on "dynamic visual computing" and successfully developed prototypes of several self-driving cars. The Institute of Automation of The National Defense University of Science and Technology has been studying the key technologies of the autonomous systems since the 1980s, and has introduced the CITAVT series of unmanned vehicles, which can complete low-speed autonomous driving on unstructured roads and autonomous driving on structured roads (Fig. 1.2).

**Fig. 1.2** CITAVT-IV autonomous car

3. **The Booming Stage**

Since 2005, some UAVs have been equipped with armaments to perform more military tasks, such as target bombing, ground attack, air combat, etc. Notwithstanding the above, the attention is now on developing invisible drones. Concurrently, drones for civilian usage are also progressing. In 2014, Amazon's delivery drone prototype, Prime Air, was launched. Prime Air is capable of delivery within a radius of about 20 km. In December 2016, Amazon completed the first commercial UAV delivery. In May 2017, the Amazon invested a new research and development centre at the outskirts of Paris, France, with the plan to provide a 30-min drone delivery service. The centre aims to develop the safest and most advanced unmanned traffic management software. At the current stage, the unmanned ships have already been developed to reach rather advanced level for military and civilian usage. Unmanned ship technology is currently moving towards commercialization and large-scale development. It is becoming increasingly possible to develop unmanned cargo ships that could travel long distance across the oceans at much greater scale that could meet higher complexity and safety requirements. In May 2016, the Institute of Atmospheric Physics, Chinese Academy of Unmanned System Vehicle Research and Development team had successfully developed an unmanned semi-submersible boat for marine meteorological usage. The unmanned boat could perform real-time data acquisition of the sea temperature, humidity, air pressure, wind speed and direction, and sea surface temperature. In 2018, a Chinese company conducted a unmanned boat testing in the South China Sea which involved a total of fifty-six 56 unmanned boats. The test was fully controlled and operated autonomously by the boat without any human intervene, so that it would not be restricted by transmission and could allow the unmanned boat to perform tasks in farther waters, thereby increasing efficiency tremendously. Moreover, in the future, such unmanned boats could also undergo large-scale production. These unmanned boats could be installed with a vast number of lethal weapons which could be used against offensive forces. In July 2018, the 6th Millitary-Civilian Integration Technology and Equipment Expo exhibited an "80-knot marine battle platform". This unmanned speed boat which could reach a speed of more than 70 knots and could perform marine resource survey, patrol, search and rescue, and active surveillance. Additionally, it is easy to maintain and has a long service life span. The ship cabin could also be used for the transportation of people and goods. It could also be equipped with artillery such as machine guns and small missile launchers. In comparison to unmanned aircrafts and boats, the unmanned vehicle development is even more advanced. In 2009, Google began its research and development of its own driverless vehicle project with the support of DARPA. That year, Google, through a modified Toyota Prius Adams, travelled 1.4 million miles at the Pacific coast which lasted for more than a year. Many engineers who had worked at DARPA between 2005 and 2007 joined Google's team. They had used video systems, radar and laser auto-navigation technology for their research. A French company, Induct, developed the Navia Shuttle which is more suitable for transportation within schools and parks in comparison with the other traditional driverless vehicles. This shuttle vehicle which could be operated at a maximum speed of 13 miles/hr can automatically learn routes and sense obstacles using the lidar sensor. In 2014, Google launched its own in house designed driverless car, and in 2015, the first prototype car was officially unveiled and tested on the public roads. In 2015, EasyMile's unmanned bus, EZ10, was put into operation in France. EZ10 adopted the GPS, cameras, radar sensors and other sensing devices to detect its position and obstacles. Passengers could easily view the current location of the unmanned bus and make a request call through the mobile application. In December of the same year, Baidu completed its driverless car testing on the designated highway reserved for autopilot testing in Beijing. In April 2016, Changan Automobile successfully completed the 2000-km driverless super test. In the same year, Ford Fusion organised the first driverless vehicle road show. In 2017, the University of Michigan launched the MCity project which built the world's largest driverless car testing site. In the same year, Baidu officially announced its strategic collaboration with Bo Shi to develop a high-resolution navigation map that could provide a more accurate real-time positioning system for autonomous vehicles. Meanwhile, at the press conference, the collaboration results between Bo Shi and Baidu to develop the enhanced version of the freeway assistance function in a pilot car were showcased. In March 2017, Tesla launched the Autopilot 8.1 system in order to greatly enhance the standards of the automated vehicles. According to statistics, Tesla has travelled more than 2.022 billion miles in autopilot mode. In August 2018, the autonomous driving technology research and development company, Torc Robotics, announced a partnership with the leading manufacturer in vehicle machinery chipset, NXP., in a joint research and development of the autopilot system. In September 2017, Qualcomm launched a new C-V2X chipset and reference design which allows the vehicle manufacturers to better deploy the fully autonomous vehicle's required communication system. In 2018, Baidu planned to cooperate with Xiamen Jinlong to take the lead in achieving a small-scale mass production and the pilot operation of the driverless minibuses. In January 2018, Toyota unveiled the e-Palette Concept car at the CES exhibition in Las Vegas, which is a commercialised fully electric car that is designed specifically for companies such as online car and delivery services. In August 2018, Japan's ZMP company, which is dedicated to the driverless technology, and the large-scale taxi company, Nippori Maru Transportation, conducted trial runs of unmanned taxis for the passengers in Tokyo. In terms of open road testing, the California Vehicle Authority officially announced that it will allow unmanned vehicles to be tested on the road from 2018 onwards (Fig. 1.3).

**Fig. 1.3** Current stage of unmanned systems

## 1.3 Key Technologies of Unmanned System

### *1.3.1 Sensor Technology*

Both humans and higher-level animals have rich sensory organs that can sense external stimuli and acquire external environmental information through various senses of the human body such as sight, hearing, taste, touch and smell. Autonomous unmanned systems can also obtain surrounding environmental information through various sensors. The sensors have an essential and important role for the robots. The understanding of the information acquired by these sensors is essentially dependent on the level of autonomy of the system and the level of learning ability. It can be said that the sensor technology fundamentally determines and yet also restricts the development of the intelligent unmanned system's environment sensing ability (Fig. 1.4).

At the current stage, a variety of sensors had been created which include a millimeter-wave radar that is used for distance measurement and imaging, the laser light radar, the infrared sensor, the visual sensor and the global positioning system which is primarily used for sensing the external environment for information.

1. Millimeter-wave radar: The millimeter-wave radar (operating wavelength of about 8 mm) was first used for airport traffic control and marine navigation for its superior characteristics such as high resolution, high precision, and small antenna aperture. However, due to technical difficulties, the development of the millimeter wave radar was limited. For instance, as the operating frequency increases, the power source output and efficiency decreases, resulting in the increase of losses in the receiver mixer and transmission line. Nonetheless, with the advancement of hardware devices and the improvement of the algorithms,

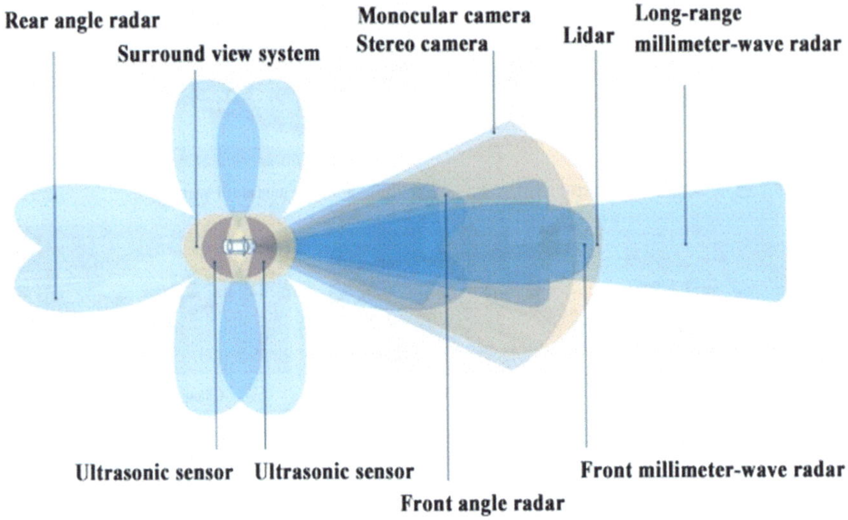


**Fig. 1.4** Unmanned vehicle sensor distribution schematic

the millimeter wave radars have become more widely used in the unmanned systems.

The millimeter wave radar is suitable for using the 30–300 GHz millimeter wave frequency domain, and possesses the ability to function without adverse impact due to fog, smoke, dust and other weather conditions (except for heavy rain). Additionally, it can obtain an accurate distance and speed measurement with reference to the obstacle. Although the millimeter wave experiences energy loss during transmission, it can still fulfil the requirement of front vehicles detection. At the same time, the millimeter-wave radar has the advantage of its small size as compared to the other sensors, and has been widely used in the ground unmanned mobility platforms. It offers not only distance measurement application, but also imaging application such as active imaging and passive imaging.

2. Lidar: The origin of the Lidar started with stereophotogrammetry technology. With the development of science and technology and the wide application of computers, the digital stereophotogrammetry gradually progressed and matured. The corresponding software and digital stereophotogrammetry work stations also started to popularize. The robot navigation applications had increased tremendously over time, and this is mainly due to the many advantages of the laser-based distance measurement technology such as its high precision in distance measurement by scanning the laser beam in two- or three-dimensional manner. The Lidar uses a relatively high frequency to provide a large amount of accurate distance information. Compared with other distance sensors, Lidar can fulfil both accuracy and speed requirements at the same time, which is especially

**Fig. 1.5** Highway lidar point clouds

suitable for mobile robots. In addition, the lidar can work not only in ambient light, but also in the dark. In fact, it can measure better in the dark (Fig. 1.5).

3. Infrared sensors: The infrared sensor has a history of 40 years since its first generation. The infrared sensor could be divided into four generations according to its characteristics:

   The first generation (1970s–80s) used single or multiple optical components to perform scan imaging; the second generation (1990s–2000s) used a 4 × 288 scan focal plane for imaging; the third generation used a new type of sensor for the gaze focal plane for imaging; and lastly the fourth generation features a multiband, smart and flexible system and chipset that could produce a large area array of high resolution recognition and high-performance digital signal processing. It also has single-chip multi-band fusion detection and recognition capabilities.

   Infrared sensing is a non-contact passive measuring sensor. The biggest feature of the infrared light is its photothermal effect and radiant energy, which makes up the largest photothermal effect region in the spectrum. Infrared light is an invisible light. Like all electromagnetic waves, it has the properties of reflection, refraction, scattering, interference, absorption and so on. The infrared sensor's night vision technology, thermal imaging technology and the ability to detect up to 400 m range makes it an indispensable sensor in the unmanned systems.

4. Vision Sensors: Vision sensor appeared in the 1950s and thereafter developed very rapidly. After the 1970s, other systems which offer more practical application such as in IC manufacturing, precision assembly of electronic products, quality inspection of beverage cans and boxes, positioning, etc., started to appear. Vision sensors are classified into monocular sensors, dual (multiple) sensors, RGB-D, etc. The monocular sensor is simple in structure and low in cost. It can be used in both indoors and outdoors. However, only a relative depth can be estimated and the absolute depth cannot be determined. The binocular

sensor can estimate both the depth during motion and at rest, but the configuration and calibration are complex. The depth range is also limited by the dual-purpose baseline and resolution. RGB-D is a kind of camera that can produce colour map and measure depth. Generally, the distance between the object and the camera is directly measured by the speed of light. The speed is fast but the visual field is small. As it has a low resolution, it is mainly used indoors.

Vision sensor is an essential sensor for the current mobile robot environment sensing technology. The machine vision technology has benefited from the development of image processing technology and recognition algorithm in the recent years, enabling the surrounding environment image to be acquired through the camera, and hence making the 3D perception model possible. Vision sensors are chosen for use in the unmanned systems due to their accuracy, ease of use, rich functionality and reasonable cost.

Vision sensing is based on machine vision to obtain image information of the vehicle's surrounding environment. It senses the surrounding environment through image processing. Vision sensor can directly obtain colour information, and the camera has the advantages of convenient installation, small size and low energy consumption. Its drawback includes susceptibility to interference light and low accuracy in measuring three-dimensional information. Laser sensing is achieved by using the radar to scan the surroundings of the vehicle to produce two-dimensional or three-dimensional distance information. Through the analysis of the distance information, the road traffic conditions could be detected. It can directly obtain the three-dimensional distance information of the object, has high measurement accuracy and is insensitive to lighting contrast changes. However, it cannot obtain the colour information of the environment, and because the radar is large in size, it is expensive and inconvenient for vehicle integration. Furthermore, the lidar measurement accuracy is easily affected by inclement weather such as rain and snow. The millimeter wave radar is small in size, light weight and has high spatial resolution. Compared with the optical lenses from infrared sensor, laser, and the television, the millimeter wave radar has a relatively strong ability to penetrate fog, smoke, dust and other inclement weather conditions (except for heavy rain) at any time. However, due to its wavelength, the millimeter wave radar's detection range is very limited and the coverage area is fan-shaped and hence has a blind spot area. The infrared system including infrared imaging can work under all weather conditions at any time of the day. Pedestrians could also be detected at night or under strong lighting conditions. Notwithstanding the above, the millimeter wave radar, lidar, and infrared sensor are unable to identify traffic signs.

### 1.3.2 Sensor Fusion Technology

Information fusion is a multidisciplinary and emerging technology. Different algorithms and models are involved in different levels of integration. At present, domestic and international research has been devoted to the research of information fusion algorithms, and some progress has been made. Some of the classic algorithms that are used in the multi-sensor information fusion technology are introduced below.

1. The Weighted Average Method

The weighted average method is the simplest and most intuitive method of information fusion, which is to average the weighted data values that are provided by multiple sensors to achieve the final fusion value. The main challenge of using the weighted average method is on how to assign the appropriate weight values to the data obtained from the sensors and how to choose a suitable weight calculation method.

2. Kalman Filtering

The Kalman filter can estimate the current value of the signal based on the estimated value of the previous signal state and the measured value of the current state, and then perform a recursive calculation. In the data level fusion, the raw data obtained from the sensor presents a relatively large error, and the Kalman filtering method can effectively reduce the error of the measured data and improve the fusion quality.

3. Bayesian Inference Method

Bayesian inference has a mathematical formula structure which makes it easy to understand, and requires only a moderate calculation time. However, Bayesian inference requires prior known probabilities and for the various variables to be independent of each other. This is difficult to satisfy in most practical applications, so the Bayesian inference method presents great limitations.

4. Statistical Decision Theory

By selecting an optimal deciding parameter benchmark to facilitate decision making can improve the fusion outcome. The loss function is one of the important parameters in statistical decision theory, and the method of selecting the loss function is one of the difficulties in statistical decision theory.

5. DS Evidence Reasoning

The biggest advantage and feature of the DS evidence reasoning method is that it can effectively describe the uncertain information. The evidence interval is divided into support interval, trust interval and rejection interval by the trust function and the suspicion function to express the uncertainty and unknown information. DS evidence reasoning has strict theoretical derivation. Compared with Bayesian and other inference methods, data can be synthesized without prior probability. At this time, the DS evidence synthesis formula can effectively fuse the measured data to obtain more accurate judgment results when the evidence provided by the measured data is not much different. However, the theory also has problems such as one-vote veto and subjective composition being too large.

6. Fuzzy Theory

Fuzzy theory is an algorithm that imitates the way that human beings think. It summarizes the common features of the objects through cognitive processing and abstract extraction. On the computer, fuzzy function is used to extract some function indicators.

The multidisciplinary nature of the fuzzy theory allows the integration of different algorithms to solve the problem of uncertainty. In the information fusion system, the fuzzy theory can effectively improve the fusion effect. However, the limitation of the fuzzy theory is on how to construct reasonable and effective membership function and index function.

7. Neural Network Method

The neural network uses distributed parallel information processing to process information through a large number of network nodes that simulate the human neuron systems. Its characteristic is that it can process parallel information at high speed, and can solve the problem of excessive information in the information fusion system. Neural networks also have the ability to handle nonlinear relationships, and the algorithms are easy to implement in computers. The main challenge of the neural network fusion algorithm is that the learning method itself still possess some problems, such as stability problems, generalization, and the lack of effective learning mechanism.

### *1.3.3 Positioning Navigation Technology*

With the rapid development of technology and globalization today, positioning and navigation technology plays an increasingly important role in our daily work and life. It is widely used in the advanced technology applications in the marine, land and air aspects. In the unmanned automated mobile system, the navigation technology mainly provides information such as direction, position, speed and time for the motion carrier. It is used to determine the geographical location of the motion carrier itself, which forms the basis and support for path planning and mission planning.

The positioning and navigation technology that is commonly used today can basically be divided into the following categories: absolute positioning, such as positioning by means of Global Positioning System (GPS); relative positioning,

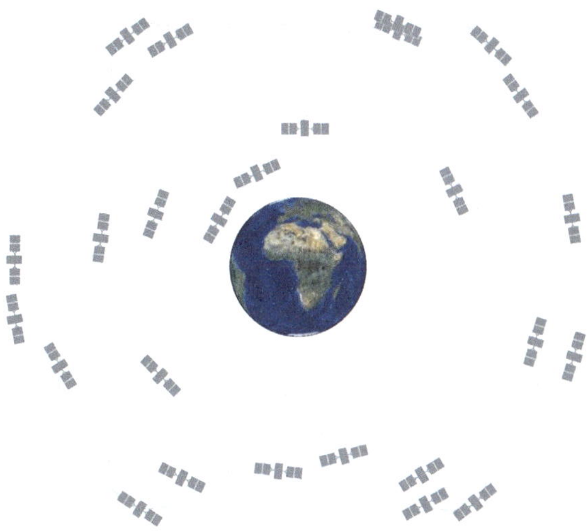

**Fig. 1.6** Global satellite system

such as odometers, gyroscopes and other inertial sensors; combined positioning; simultaneous positioning and map creation (SLAM).

The Global Navigation Satellite System (GNSS) has become an important player in the navigation field due to its globalised, weather-proof and real-time nature. At the same time, it also has shortcomings that cannot be ignored, such as poor autonomy and reliability, and is susceptible to influence by various factors (Fig. 1.6).

Inertial navigation technology is widely used in various fields. It has a strong independence and reliability. Inertial navigation has higher accuracy in a short period of time, but as the inertial navigation system (INS) drifts gradually, errors start to accumulate. Thus the work accuracy is not high for long work time, which greatly limits the application of the inertial navigation.

Since each single-navigation system has its own limitations, integrated navigation has become a research hotspot. The combination of the inertial navigation and satellite navigation is a technology with broad prospects and is currently under development.

Simultaneous Localization and Mapping (SLAM) technology related basic theories and solutions provides great scientific and application value in realizing the unmanned automated mobile systems.

### *1.3.4 Path Planning Technique*

Path planning refers to an unmanned system such as a robot seeking an effective path from the initial state target in an environment with obstacles, which can be regarded as a discrete graph search problem in continuous space. Path planning algorithms have high requirements of real-time and robustness, and efficient path planning algorithms are crucial. It is mainly divided into two parts: the construction of the environment and the shortest and optimal path search. Common path planning algorithm can be divided into two categories: strategic planning and local planning.

Conventional path planning algorithms include the annealing algorithm, artificial potential field method, fuzzy logic algorithms, and so on. Traditional algorithms often have difficulty in building the model when solving practical problems. On the other hand, graphics planning method provides the basic method for modeling. However, the methods of graphics generally have insufficient search capabilities, and thus a special search algorithm should be combined. Commonly used graphics planning methods include C space method, grid method, free space method, voronoi diagram, and so on.

When dealing with path planning problems under complex dynamic environment information, revelations from nature often play a very good role. Intelligent bionics algorithms are algorithms discovered by people through bionics research, such as ant colony algorithm, neural network algorithm, Genetic algorithms, etc. Other algorithms mainly refer to algorithms that have strong search capabilities or can function well in discrete path topology networks, such as A* algorithm, Dijkstra algorithm, floyd algorithm, and so on.

## Bibliography

1. Zijun Yang, Fengying Liao, Qing Wei (2005) Application of GPS in mobile robot navigation and positioning system. Ship Electron Eng 25(6):5–7
2. Xu W (2006) Research on robot positioning/navigation technology based on laser radar environment information processing. Master degree thesis of Nanjing University of Science and Technology, 6
3. Zhang M, Liu P, Xu Y, Wang J, Liu Y (2006) Research on key technologies of environmental awareness of ground unmanned vehicles. In: National testing and fault diagnosis technology seminar
4. Yang H, Qian K, Dai X, Ma X, Fang F (2013) Based on kinect to create three-dimensional map of a mobile robot indoor environmental sensor. In: 2013 China's intelligent automation academic conference proceedings (volume III)
5. Zhu Z (2014) Unstructured scene understanding based on 3D data for unmanned vehicle navigation. Zhejiang University PhD thesis, Zhejiang
6. Luo D, Qiu T, Lu Q (2006) Ultrasonic technology and applications (IV)—application of ultrasonic technology in other aspects. Daily Chem Ind 36(2):120–123
7. Wang D, Wang J (2013) A research overview on environmental perception technology of mobile robots in unknown environment. Mach Tool Hydraulics 41(15):187–191
8. Yang J, Zhang M, Shang Y (2008) A method for visual image feature extraction and segmentation of mobile robots. Robotics 30(4):311–317
9. Sural S, Qian G, Pramanik S (2002) Segmentation and histogram generation using the HSV color space for image retrieval. In: Proceedings of the 2002 international conference on image processing. Piscataway, NJ, USA. IEEE, pp 589–592

10. Ulrichi N (2000) Appearance-based place recognition for topological localization. In: Proceedings of the IEEE international conference on robotics and automation, San Francisco: IEEE Press, pp 1023–1029
11. Shi C, Hong B, Zhou T et al (2007) Topological map creation and navigation in large-scale environments. Robotics 29(5):433–438
12. Zixing Cai, Hangen He, Hong Chen (2002) Problems in the research of mobile robot navigation control in unknown environment. Control Decision 17(4):385–464
13. THRUNS (2002) Robotic mapping: a survey. Pittsburgh: CMUCS02111, School of Computer Science, Carnegie Mellon University
14. Feng L (2008) Research on navigation and positioning technology of mobile robot based on multi-sensor information fusion. Southwest Jiaotong University, Chengdu
15. Kang J (2013) Research on key technologies based on multi-sensor information fusion. Harbin Engineering University, Harbin
16. Hyun-Joeong L, Moon Sik K, Min Cheol L (2007) Technique to correct the localization error of the mobile robot positioning system using an RFID. In: SICE, annual conference, pp 1506–1511
17. Heesung C, Christiand (2010) Autonomous navigation of mobile robot based on DGPS/INS sensor fusion by EKF in semi-outdoor structured environment. In: The 2010 IEEE/RSJ international conference on intelligent robots and systems, Taipei, Taiwan, pp 1222–1227
18. Sommerkd, Kuehno, Leonfp, et al (2009) A Bayesian approach to information fusion for evaluating the measurement uncertainty. Robot Auton Syst 57:339–344
19. Basiro O, Yuan X (2007) Engine fault diagnosis based on multi-sensor information fusion using dempster—Shafer evidence theory. Inf Fusion (8):379–386
20. Kun Wang, Minzhou Luo, Jianghai Zhao (2010) A research overview on information acquisition methods in indoor mobile robot navigation. Rob Appl 22(2):38–42
21. Ling Zhao (2007) Research on mobile robot navigation system based on fusion of vision and ultrasonic sensors. Wuhan University of Technology, Wuhan
22. Zhao X, Luo Q (2008) Survey on robot multi-sensor information fusion technology. In: Proceedings of the 7th world congress on intelligent control and automation, Chongqing, China, pp 5019–5023
23. Xiufeng Tao, Mingzheng Zhou (2006) Multi-sensor information fusion algorithm based on support vector machine. Comput Technol Develop 16(6):177–183
24. Xiaogang Jian, Hongsheng Jia, Laide Shi (2009) Research progress of multi-sensor information fusion technology. Chin J Mech Eng 7(2):227–232

# Chapter 2
# Millimeter Wave Radar Technology

## 2.1 Introduction

Radar is the abbreviation of 'Radio Detection and Ranging'. It means to detect and estimate distances via the electromagnetic waves. The millimeter wave is an electromagnetic wave that lies between the infrared light wave and the microwave frequency band. Different classification methods are applied on the different classes of the millimeter wave radar. The millimeter wave radar could be classified into pulse radar and continuous wave radar based on the nature of the radar working system. The working mode of the continuous wave radar is further classified into constant frequency continuous wave (CW), frequency shift keying (FSK), phase shift keying (PSK), frequency modulated continuous wave (FMCW) and other methods. Based on platform classification, it can be categorized into vehicle millimeter wave radar, shipboard millimeter wave radar, airborne millimeter wave radar, spaceborne millimeter wave radar, etc. According to the different requirements of radar ranging, millimeter-wave radar can be classified into long-range radar (LRR), medium-range radar (MRR) and short-range radar (SRR). Additionally, based on field of application, millimeter-wave radar can be classified into guided radar, fire control radar, target detection radar, millimeter-wave earth observation radar, millimeter wave proximity detection radar, etc.

In the recent years, as millimeter-wave devices advance, technologies such as circuit design technology and antenna technology also mature tremendously over time. Millimeter-wave radar technology, in relevant fields such as unmanned system environment sensing, radar detection, missile guidance, satellite remote sensing, electronic countermeasures, etc., has also progressed with increased application. Given millimeter-wave radar's high frequency, short wavelength, wide frequency band, small size, light weight, good concealment and maneuverability, it plays an important role in the military and defense construction, and is also widely adopted in the national economic construction. The millimeter-wave radar has the ability to penetrate haze, fog and mist and performs much better than the



X. Bi, *Environmental Perception Technology for Unmanned Systems*, Unmanned System Technologies, https://doi.org/10.1007/978-981-15-8093-2_2

centimeter-wave radar as it has anti-noise interference capability and multi-path resistance which can fully utilize its precision detection and precision tracking. Hence it had been developed as a military project by major developed countries. In the civilian aspect, the millimeter-wave is a better substitute to X-ray to achieve better image detection results of objects that are hidden underneath the garments, and thereby strengthen the security. For meteorological detection, the millimeter-wave radar which has a high spatial and temporal resolution could reflect the cloud vertical and horizontal structure more accurately. Hence, it is more suitable for monitoring cloud changes than ordinary weather radars, making it an important tool for three-dimensional fine structure cloud detection. At the same time, millimeter-wave radar is an indispensable sensor used for environmental sensing in the intelligent driving system that provides the functions of anti-collision, automated parking, pedestrian detection and many others. This implies great significance in reducing traffic incidents and protecting the lives and properties of pedestrians and passengers.

## 2.2 Millimeter-Wave Radar Concept and Characteristics

Millimeter-wave radar is a detection radar that operates using the millimeter wave. It is an electromagnetic wave that lies in the spectrum between the microwave and the light wave with a frequency range between 30–300 GHz and wavelength between 1–10 mm. The IEEE had issued the 30–300 GHz frequency band as the standard frequency range of the millimeter wave, and named the frequency band of 27–40 GHz as the Ka band, 40–60 GHz frequency band as the U band, 75–110 GHz frequency band as the W band, 110–170 GHz as the D band. The Ka-band has the characteristics of wide bandwidth, reduced interference, and small device size. Hence, Ka-band satellite communication systems could provide as a novel approach for high-speed satellite communications, gigabit-class broadband digital transmission, high definition television (HDTV), satellite news gathering (SNG), VSAT services, direct-to-home (DTH) services, and personal satellite communications. The disadvantage of the Ka band is that the rain attenuation is large, and the requirements for the device and the process are extensive. In the Ka-band tone, the antenna size of the Ka user terminal is not primarily limited by the antenna gain, but is limited by the ability to suppress interference from other systems (Figs. 2.1 and 2.2).

In comparison with the microwave, millimeter wave wavelength is short and has high frequency, wide bandwidth and large Doppler shift. The short wavelength results in the small electronic component packaging size and systems to be compacted, and hence, is light weight. The manufacturing precision requirement is strict and hence results in high cost. Additionally, the millimeter wave achieves a narrow beam of high gain more easily for the same diameter antenna. This enhances the multi-radar system's target recognition ability and results in small ground clutter interference and multipath effect. In addition, the linearity of propagation is good,

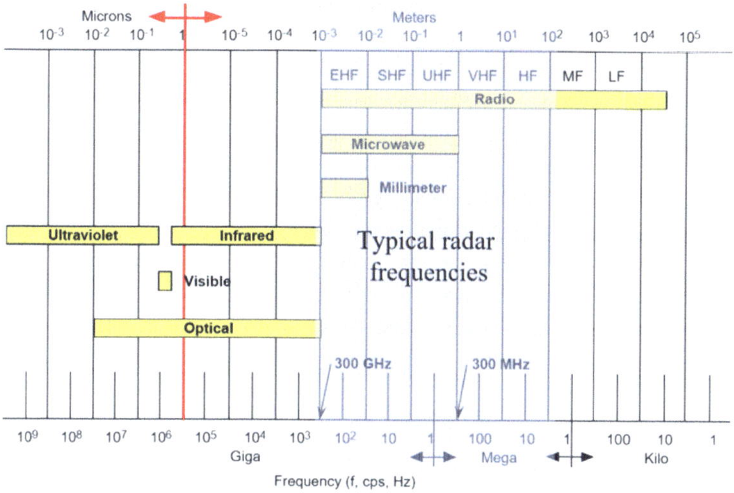


**Fig. 2.1** Electromagnetic wave band

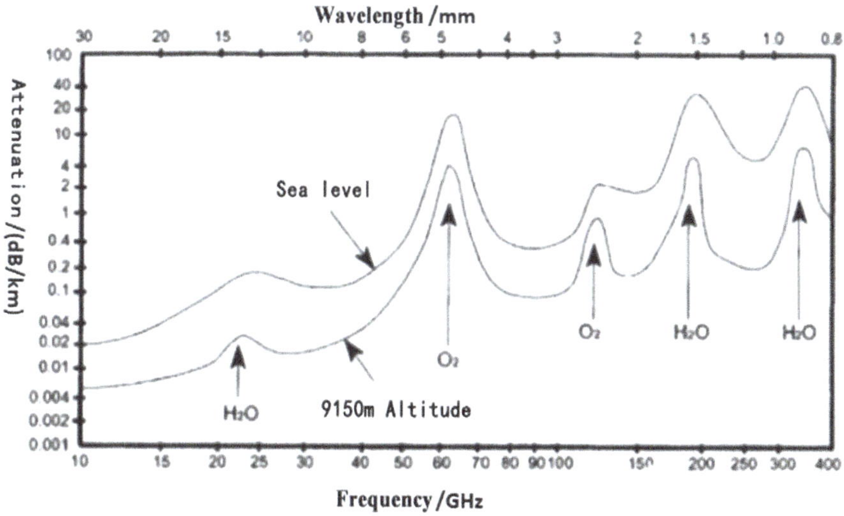


**Fig. 2.2** Atmospheric attenuation trend of different frequency bands of millimeter wave

the diffraction is small and the interference to other communication devices can be reduced. The available bandwidth is wide which can improve the information transmission rate. In the radar system, narrow pulse or wideband FM signals can be used to study the fine characteristics of the target. As the transmission capacity is large in the communication system, the distance resolution of the precision tracking radar and the target recognition radar is improved. The broadband spread spectrum capability can be used to suppress multipath effects and the ground clutter. Additionally, the rich spectral resources and large Doppler shift can improve the detection and recognition ability of low-speed moving objects. In comparison with infrared and visible light, its signal transmission is less affected by harsh weather conditions and dust, smoke and other site conditions, and can work around the clock. Current application research focuses on several window frequencies, such as 35, 45, 94, 140, and 220 GHz, which can achieve medium-range multi-channel communication and TV image transmission. With its high transmission rate, it is conducive to low-intercept probability communication such as spread spectrum communication and frequency hopping communication. At the same time, it has three absorption peaks (60, 120, 200 GHz), which can be used to achieve close-range radar, secure communication and satellite communication.

## 2.3 Radar Equation

When the electromagnetic wave that is emitted by the radar touches the target, it will be reflected. The power transmitted is related to the size, orientation, physical shape and the material of the target. These factors are combined and represented as the radar cross-sectional area (RCS), $\sigma$, which is expressed as the ratio of the target transmit power to the target receiving power:

$$\sigma = \frac{P_r}{P_D}\left(m^2\right) \tag{2.1}$$

The reflected power of the target, $P_r$, is also radiated to the surroundings. If the effective receiving aperture of the antenna is $A_e$, the magnitude of the reflected power received by the antenna can be expressed as:

$$P_{Dr} = \frac{P_t G \sigma A_e}{\left(4\pi R^2\right)^2} \tag{2.2}$$

And because the effective receiving aperture of the antenna, $A_e$, has the following relationship with the antenna gain:

$$G = \frac{4\pi A_e}{\lambda^2} \tag{2.3}$$

The antenna that receives the target reflected power can be rewritten as:

$$P_{Dr} = \frac{P_t G^2 \sigma \lambda^2}{(4\pi)^3 R^4} \tag{2.4}$$

The above formula shows the relationship between the target distance, $R$, and the power received by the radar from the target, $P_{Dr}$. Assuming that the radar has a minimum detectable signal power, $S_{min}$, the farthest detection distance of the radar that can be reached is expressed as:

$$R_{max} = [\frac{P_t G^2 \sigma \lambda^2}{(4\pi)^3 S_{min}}]^{1/4} \tag{2.5}$$

It can be seen that for the farthest detection distance to be doubled, under normal conditions where the remaining parameters are unchanged, the radar's transmission power needs to be increased by 16 times. Therefore, showing that the radar's detection range is closely related to the radar's transmit power.

## 2.4 Millimeter-Wave Radar Analysis

Generally, the radar system will actively emit an electromagnetic wave signal. The electromagnetic wave signal then forms a reflected echo after encountering an object. Using the information of the reflected echo, the radar can detect the target and measure its coordinates. According to the different working systems of the radar, it can be divided into the pulse system radar and the continuous wave system radar. The pulse system radar intermittently emits the electromagnetic waves, and the intermittent wave radar using the transmitted waveform will receive the target echo. On the other hand, the continuous wave system radar continuously emits the electromagnetic waves, and the radar receives the echo of the target while the electromagnetic wave is emitted.

### *2.4.1 Pulse System*

The pulsed radar periodically emits a waveform. The waveform emission period is called a pulse reception interval (PRI). In a pulse repetition period, the ratio of the electromagnetic wave signal emission time to the pulse repetition period is called the duty ratio. For pulsed radars, the duty cycle is usually less than 20% due to the actual device level. The size of the PRI determines the fuzzy-free distance range of the pulsed radar. The larger the PRI, the larger the fuzzy ranging distance.

The pulse signal can be divided into three forms, a non-coherent pulse signal, a coherent pulse signal, and a parametric variable phase coherent pulse signal.

Coherent pulse means that the initial phase between the pulses is deterministic, that is, the initial phase of the first pulse may be random but the phase between the subsequent pulse and the first pulse is deterministic, which forms the basis for the extraction of the Doppler information group. Non-coherent pulse means that the initial phase between the pulses is random and uncorrelated with each other. Therefore, radars using non-coherent signals can only use the amplitude of the target echo to detect the target, which limits the performance of the radar. At present, most of the radar signal processing utilizes the amplitude and phase of the signal. The radar adopts a full-coherence system, and the transmitted signal is a coherent pulse signal.

The distance ranging principle of the pulse millimeter-wave radar is conducted by transmitting a single or a series of very narrow electromagnetic wave pulses to the target, and then measure the time taken for the emitted electromagnetic wave to travel to the target and back to the receiver $\Delta t$. The distance of the target, $R$, is then calculated based on the time measured (Figs. 2.3).

Target distance:

$$R = \frac{c\Delta t}{2} \tag{2.6}$$

In the above radar ranging principle, there is a delay between the received echo and the transmitted waveform. Due to the periodicity, for the echo 2, it is impossible to judge whether the pulse 1 or the pulse 2 is generated, thereby generating the distance fuzziness. The fuzzy target distance problem can be solved by the variable-cycle pulse signal.

The parametric variable period pulse signal is a pulse sequence with a pulse repetition period of $(T_1, T_2, \ldots, T_L)$, which is also called a L-staggered pulse sequence, as illustrated in Fig. 2.4.

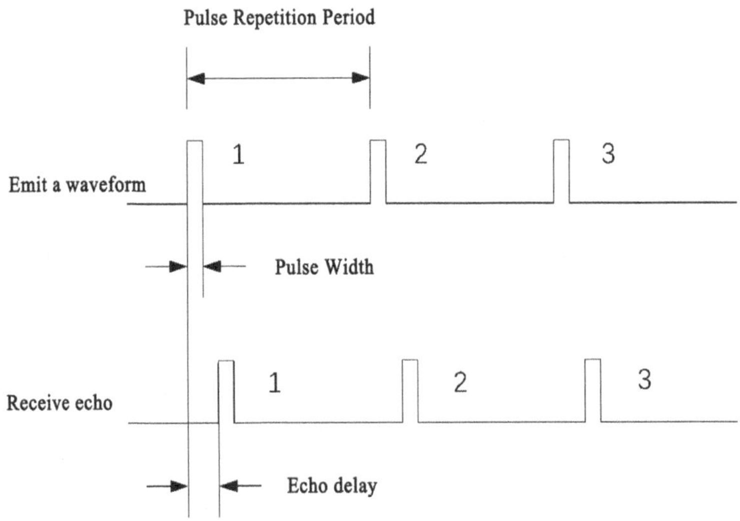


**Fig. 2.3** Radar ranging principle

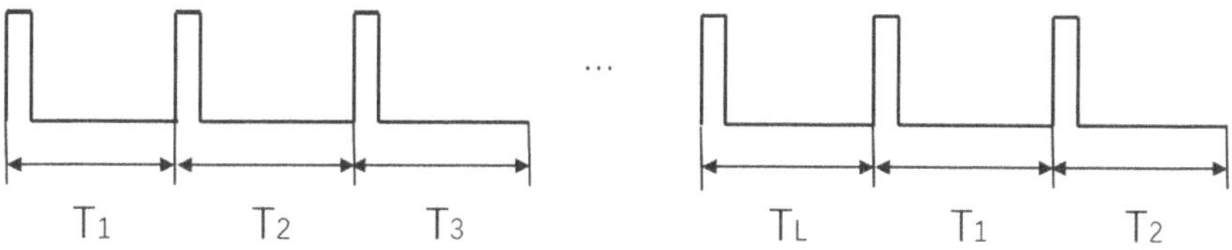


**Fig. 2.4** Parametric variable period pulse signal

Assuming $T_1 = K_1 \Delta T, T_2 = K_2 \Delta T, \ldots, T_L = K_L \Delta T$ and $K_1, K_2, \ldots, K_L$ as the mixed code, the distance and velocity are solved by analyzing the fuzzy function graph of the variable-period pulse signal.

Usually, the pulsed radar transmitting antenna and the receiving antenna are shared. During the time of transmitting the waveform the radar does not receive the echo, but receives the echo during the transmitting interval and then processes it. Since the radar cannot receive signals during the transmit waveform, the radar has a certain blind distance. Furthermore, the millimeter-wave radar using the pulse method needs to transmit a high-power pulse signal in a short time, and the over-pulse signal which controls the pressure-controlled oscillator of the radar instantaneously jump from a low frequency to a high frequency. Additionally, before the amplification of the echo signal is performed, it needs to be strictly isolated from the transmitted signal. Also, there are problems such as the achievability of the narrow pulse waveform and the switching time of the transceiver. Therefore, it is difficult for the pulsed radar to detect the near-distance target. For example, it is difficult to achieve the target within a 100 m probe.

The pulsed radar uses a time difference between the transmitted waveform of the PRI and the received waveform for ranging processing. The range resolution is related to the transmit pulse width. The higher the range resolution, the narrower the transmit pulse width is required. For example, if the distance resolution is 1 m, the transmit pulse width is required to be 6.67 ns and the receiver bandwidth is 150 MHz. Generating the 6.67 ns narrow pulse is difficult to achieve, and the 150 MHz bandwidth of the receiver means that the system also requires a higher A/D sampling rate, and hence higher cost of the engineering implementation.

### *2.4.2 Pulse Compression*

Pulse compression refers to the process of emitting wide pulses and then process the echo pulses to obtain a narrow pulse. Hence, the pulse compression radar could achieve both the objectives of keeping the narrow pulse at high range resolution while also maintaining the strong detection capability of a wide pulse. On the other hand, pulse compression systems are complex in structure and produce side lobes.

The implementation of pulse compression requires the following:

(1) The transmitted pulse must have a nonlinear phase spectrum, or the product of its pulse width and effective spectral width must be greater than 1;
(2) The receiver must have a compression network whose phase-frequency characteristics should be phase-conjugated with the transmitted signal.

Based on the above requirements, an ideal pulse compression system can be constructed as shown in Fig. 2.5.

In the ideal pulse compression system model, it is assumed that the signal is not distorted and the gain is 1 during the process of the wave propagation and the target emission, as well as during the transmission process in the microwave channel, transmitting and receiving antennas and compression network. Hence, the transmitted pulse signal is also the target echo pulse signal at the input of the receiver compression network. Its envelope width is given as $\tau$, and the spectrum is given as:

$$U_i(\omega) = |U_i(\omega)|e^{j\phi_i(\omega)} \tag{2.7}$$

The frequency characteristic of the compressed network is given as $H(\omega)$, and the formula is given as:

$$H(\omega) = K|U_i(\omega)|e^{-j\phi_i(\omega)}e^{-j2\pi ft_{d_0}} \tag{2.8}$$

In the formula, K is a proportional constant, which normalizes the amplitude-frequency characteristics and is a fixed delay of the compression network. After compression, the envelope width of the output signal is compressed as $\tau_0$, and the output expression of the pulse compression is given as:

$$U_0(\omega) = K|U_i(\omega)|^2e^{-j2\pi ft_{d_0}} \tag{2.9}$$

There are two ways to describe the operation of a pulse compression radar. One is to modulate a wide pulse to increase its bandwidth based on the fuzzy function. The wide pulse signal modulated at the time of reception passes through the matched filter. The distance resolution can be obtained by analyzing the fuzzy

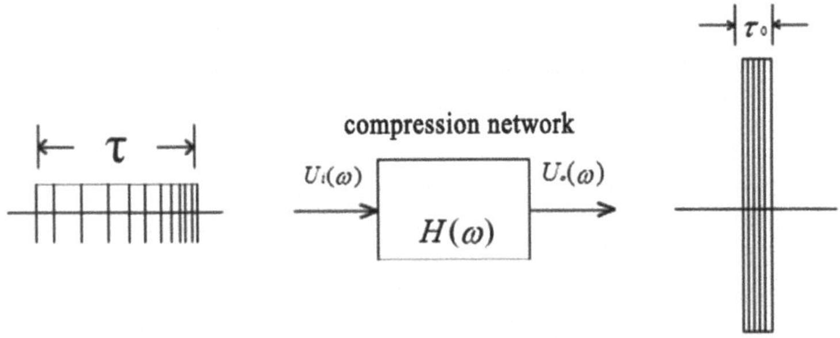


**Fig. 2.5** Ideal pulse compression system

image. A constant amplitude chirp signal is an example of a widely used pulse compression waveform, as shown in Fig. 2.6.

Another method of describing pulse compression is the chirped pulse compression as shown in Fig. 2.7. Modulating a wide pulse can be thought of as setting different "flags" in phase or frequency along different parts of the pulse. For example, the change in frequency of the chirp signal is distributed along the pulse such that each segment of the pulse corresponds to a different frequency. The modulated pulse will go through a dispersive delay line. The delay time of the dispersive delay line is a function of the frequency. The pulses of each segment would experience different delay. Hence, the decreasing edge of the pulses could be accelerated while the rising edge could be decelerated to complete the pulse compression.

### 2.4.3 Continuous Wave System

The continuous wave system could be categorized into the single frequency continuous wave, frequency-shift keying, multi-value digital frequency modulation, frequency-modulated continuous wave and chirped sequence.

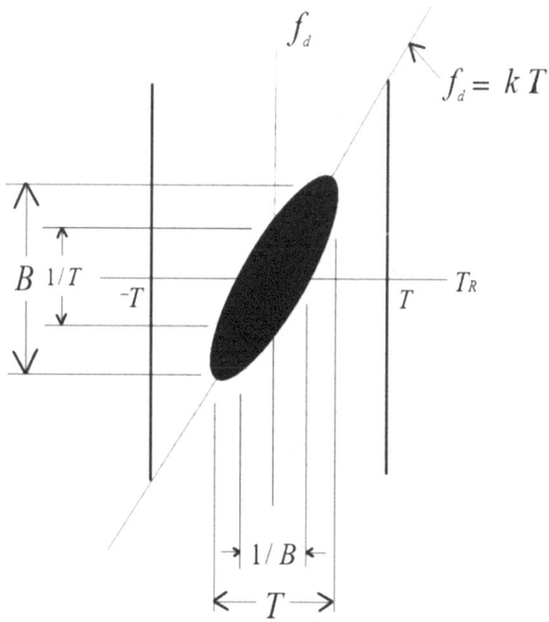


**Fig. 2.6** A two-dimensional fuzzy image of a single linear frequency modulated pulse with a width T and a bandwidth B

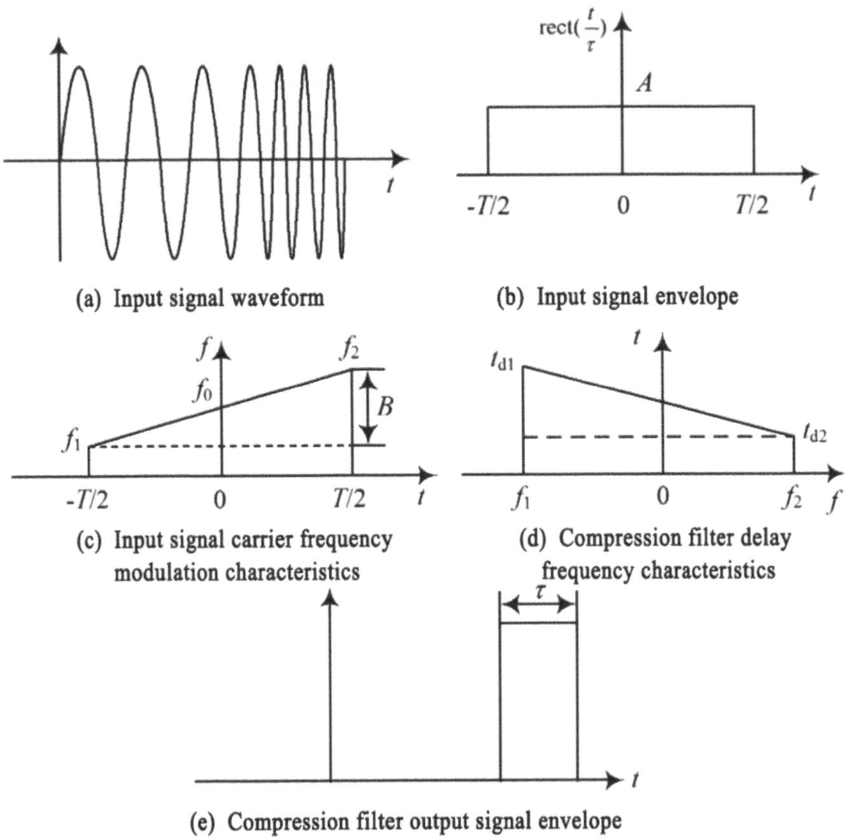


**Fig. 2.7** Chirp pulse compression

1. **Frequency Modulated Continuous Wave**

Frequency Modulated Continuous Wave (FMCW) millimeter-wave radar uses the frequency modulation to transmit signals. The commonly adopted frequency modulated wave includes triangular wave, sawtooth wave and the sine wave. When a triangular or sawtooth wave is used for frequency modulation, this is termed as the linear frequency modulated continuous wave (LFMCW).

Analysis is conducted on the situation when the target object is in a relative motion. When the received signal is reflected from the moving target, the echo signal includes a Doppler shift that is due to the motion of the target, $f_d$. The intermediate frequency signal between the rising and falling edge of the triangular wave can be illustrated in Fig. 2.8.

FMCW radar moving target echo signal could be expressed as:

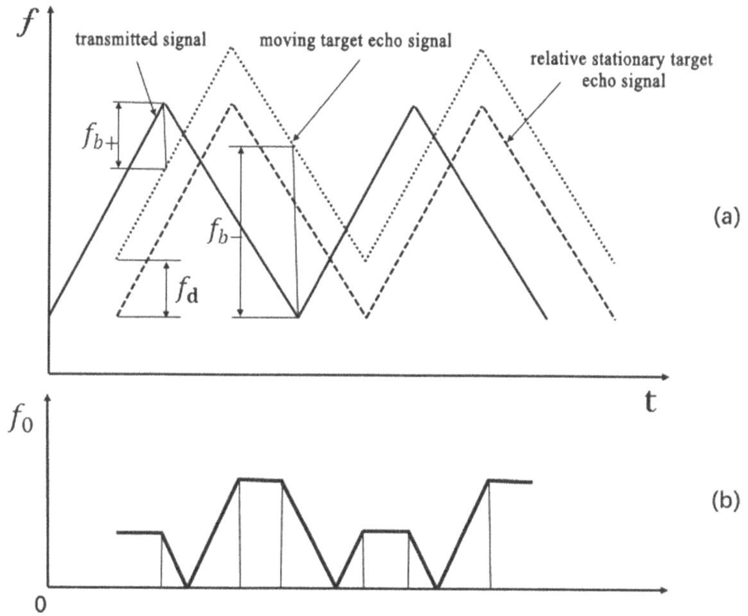


**Fig. 2.8** FMCW radar moving target echo signal

$$\begin{aligned} f_{b+} &= f_0 - f_d \\ f_{b-} &= f_0 + f_d \end{aligned} \tag{2.10}$$

In the above two equations, $f_0$ is the frequency of the intermediate frequency signal when the target object is relatively at stationary and $f_d$ is the Doppler frequency shift. Hence, $f_0$ can be expressed as follow:

$$f_0 = \frac{4\Delta FR}{cT} \tag{2.11}$$

According to the Doppler principle, the Doppler shift $f_0$ can be obtained as follow:

$$f_d = \frac{2fv}{c} \tag{2.12}$$

In the above equation, f is the centre frequency of the transmitted signal and v is the relative motion speed of the target. The sign of v is determined by the direction of the relative motion of the target. Generally, v is a positive value when the target is close to the radar system and negative when otherwise. The above equation could be further expressed as follow:

$$R = \frac{cT(f_{b-} + f_{b+})}{8\Delta F}$$
$$v = \frac{c(f_{b-} - f_{b+})}{4f} \tag{2.13}$$

Although the above two equations are derived from the relative motion of the target, they are equally applicable to relative stationary targets. Therefore, in practical applications, regardless of whether the target is in relative motion or relative stationary, the distance and velocity of the target could be calculated via the two equations by only obtaining the frequency of the intermediate frequency signal between the rising and falling edge of the modulated triangular wave. This forms the principle of millimeter-wave radar ranging.

2. **Single Frequency Continuous Wave**

Single-frequency continuous wave (CW) system uses the Doppler shift of the target echo signal to measure speed. The Doppler shift phenomenon occurs under the condition where there exists relative motion between different objects. When one of the objects emits electromagnetic wave of a certain frequency, the frequency of the received electromagnetic wave is shifted due to the relative velocity. The formula of the Doppler shift due to relative motion is given as:

$$f_d = \frac{2v_r f_0}{c} = \frac{2v_a \cos\theta f_0}{c} = \frac{2\cos\theta_e \cos\theta_a f_0}{c} v_a \tag{2.14}$$

In turn, the speed of the target object relative to the radar sensor, $v_a$, could be computed by the formula as follow:

$$v_a = \frac{c}{2f_0 \cos\theta_e \cos\theta_a} f_d \tag{2.15}$$

Wherein, $\theta_e$ and $\theta_a$ are the pitch and level angle between the center position of the radar sensor and the equivalent center position of the object respectively; and $f_0$ is the reference frequency.

It can be known from Eqs. (2.14) and (2.15) that when the radar system operates in single-frequency Doppler mode, it can only be used to measure the velocity of the objects that experience relative motion with respect to the radar system. However, it cannot obtain the distance information.

3. **Frequency Shift Keying Continuous Wave**

Frequency Shift Keying (FSK) system could make use of the phase difference of two concurrently received waves to measure the distance. At the same time, the Doppler shift could be used to measure the velocity. The target object's distance, R, and velocity, $v_a$, under this mode are given as:

$$R = -\frac{C \cdot \Delta\varphi}{4\pi \cdot f_{step}} \tag{2.16}$$

$$v_a = \frac{c \cdot f_d}{2f_0 \cos\theta_e \cos\theta_a} \tag{2.17}$$

where $\Delta\varphi$ is the phase difference; $f_{step}$ is the frequency interval; and $f_0$ is the transmission frequency. The frequency interval, $f_{step}$, and the frequency resolution of the reference frequency source, also known as the minimum frequency interval, are significantly related. It is also an important parameter index of the frequency source. If the frequency resolution of the frequency source is too large, it will cause the distance measurement to be restricted under the frequency shift keying mode. In addition, if the frequency interval is unstable, this will cause $f_{step}$ to be inaccurate and significantly reduce the accuracy of the measurement.

The frequency shift keying system is effective only for moving target. It can simultaneously detect the distance and velocity information of moving objects. For a single moving target situation, the measurement accuracy is very high. In the case of multiple moving targets, the mode detecting effect is only average.

4. **Multi-Digit Digital Frequency Modulation**

Multi-digit digital frequency modulation (MFSK) is a generalization of the FSK method, which uses different carrier frequencies to represent different kinds of digital information. MFSK transmission and reception waveforms are illustrated in Fig. 2.9. One cycle of transmitted signal includes A and B which are two mutually alternating and stepwise increasing linear modulated signal. Figure 2.9 displays solid lines which represented the transmitted signal and dashed lines as the received signal. Of which, $T_{CPI}$ is the transmit signal cycle of the multiple frequency shift keying signal which is given as $T_{CPI} = N \times T_{step}$, where $N$ is the number of steps. $f_{step}$ is the step frequency value for each frequency shift keying; $f_{shift}$ is the frequency difference between the two waveforms, A and B; $B_{SW}$ is the overall bandwidth of the waveform; $f_B$ is the frequency difference; and $f_0$ is the base band frequency.

The waveform measurement radar is theoretically derived and analyzed according to MFSK waveform characteristics.

Set $N$ number of frequencies to transmit the signal:

$$x_T(i) = A_T \exp\{-j2\pi(f_0 + if_{step})t\} \quad i = 1, 2, 3, \ldots, N-1 \tag{2.18}$$

Received signal is given as:

$$x_R(i) = A_R \exp\{-j2\pi(f_0 + if_{step})t \cdot (t - \tau(i))\} \quad i = 1, 2, 3, \ldots, N-1 \tag{2.19}$$

The intermediate frequency signal is simplified based on the principle of the mixer theory where the transmitted and received signals are mixed, and is given as:

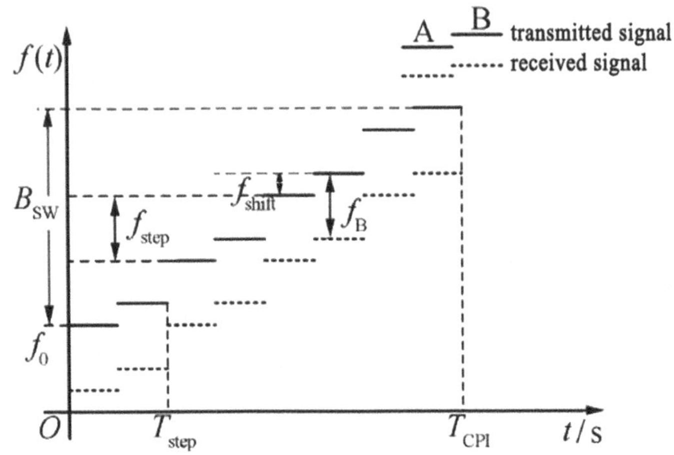


**Fig. 2.9** MFSK transmitted and received waveform

$$x_S(i) = A_S \exp\{-j2\pi(f_0 + if_{step})t \cdot \tau(i)\} \qquad i = 1, 2, 3, \ldots, N-1 \tag{2.20}$$

Wherein, $\tau(i) = \frac{2(R + viT_{step}/2)}{C}$ facilitates the FFT transformation on $N$ number of intermediate frequencies, and could further be obtained as:

$$X(k) = A_s \exp(-j\frac{4\pi}{C}f_0 R) \cdot \exp\left\{-j2\pi\frac{N-1}{2}\left[\frac{vf_0T_{step} + 2Rf_{step}}{C} + \frac{k}{N}\right]\right\}$$
$$\sum_{i=0}^{N-1}\exp(\frac{-j4\pi vi^2 f_{step}T_{step}}{2C}) \cdot \frac{\sin\left\{\pi N\left[\frac{vf_0T_{step}+2Rf_{step}}{C} + \frac{k}{N}\right]\right\}}{\sin\left\{\pi\left[\frac{vf_0T_{step}+2Rf_{step}}{C} + \frac{k}{N}\right]\right\}}. \tag{2.21}$$

From the above equation, when $k = -\frac{N(vf_0T_{step} + 2Rf_{step})}{C}$ and $X(k)$ achieves the peak, the intermediate frequency could be obtained as:

$$f_B = \frac{k}{T_{CPI}} = -\frac{2 \cdot v}{\lambda} - \frac{2Rf_{step}}{C \cdot T_{CPI}} \tag{2.22}$$

$$\Delta\varphi = -\frac{\pi \cdot v}{(N-1) \cdot \Delta v} - 4\pi R \cdot \frac{f_{step}}{C}. \tag{2.23}$$

By only finding $f_B$ and $\Delta\varphi$, using the two equations, the target distance and speed could be found.

Based on the principle of the MFSK radar measurement system, it is noted that $f_B$and$\Delta\varphi$ which are used to find the target distance and speed is a one-to-one relationship where cross aliasing does not occur. Hence, MFSK system can effectively avoid the occurrence of false targets.

5. **Chirp Sequence**

Figure 2.10 displays a chirp sequence which is composed by L number of consecutive waveforms of the same frequency modulation. Since the frequency modulation period of each sequence segment, $T_{chirt}$, is small, the received baseband signal of its single frequency modulation sequence has a large bandwidth. The frequency difference, $f_B$, can be obtained by FFT for a single FM sequence.

$$f_B = f_R - f_D = \frac{2 \cdot v}{\lambda} - \frac{2Rf_{sw}}{c \cdot T_{chirp}} \tag{2.24}$$

Wherein, it includes the target distance, $R$, and Doppler frequency, $f_D$.

The frequency difference, $f_B$, Doppler frequency, $f_D$ and phase, $\phi$ are also included in the falling frequency band of the received signal. This continuous baseband signal can be represented by the following equation:

$$S(t, l) = \exp\left(j2\pi\left(f_B \cdot t - f_D \cdot lT_{chirp} + \phi\right)\right). \tag{2.25}$$

The baseband signal, after going through sampling and independent frequency modulated sequence FFT, could be classified into $K$ number of adjacent range gates.

$$S(m, l) = \sum_{i=0}^{K-1} s(k, l) \cdot \exp\left(-j2\pi\frac{k \cdot m}{K}\right). \tag{2.26}$$

where $K$ is the sampling number of the discrete baseband signal and $m$ is the frequency difference serial number. Each individual frequency modulated sequence

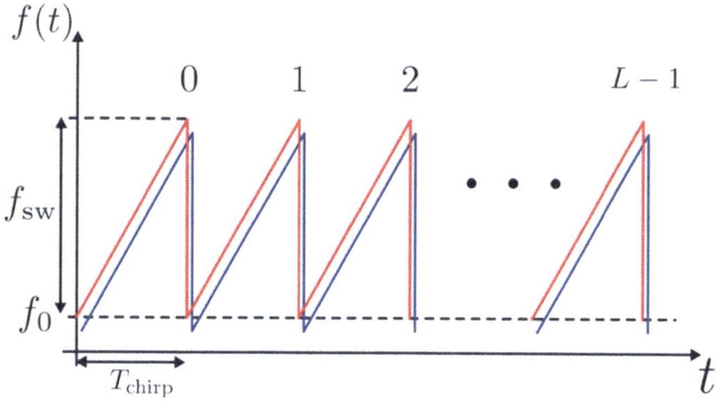


**Fig. 2.10** Chirp sequence

of the received signal contains frequency difference, $f_B$ which performs the same FFT for L number of frequency-modulated sequence. The processed data will be placed in a two-dimensional matrix as shown in Fig. 2.11.

The Doppler frequency, $f_D$, could be obtained by performing the FFT for second time via each individual range gate, that is, to perform FFT for each row in the two-dimensional matrix. The length of the FFT is given as L.

$$Q(m,l) = \sum_{i=0}^{K-1} s(m,l) \cdot \exp\left(-j2\pi \frac{l \cdot n}{L}\right) \tag{2.27}$$

where $n$ represents the sequence number of the discrete Doppler frequency.

By knowing the difference frequency, $f_B$, and Doppler frequency, $f_D$, of different targets, the distance and velocity information of the target can be obtained by the following equations:

$$\begin{aligned} R &= -(f_B + f_D) \cdot \frac{T_{chirp}}{f_{sw}} \frac{c}{2} \\ v &= -f_D \frac{\lambda}{2} \end{aligned} \tag{2.28}$$

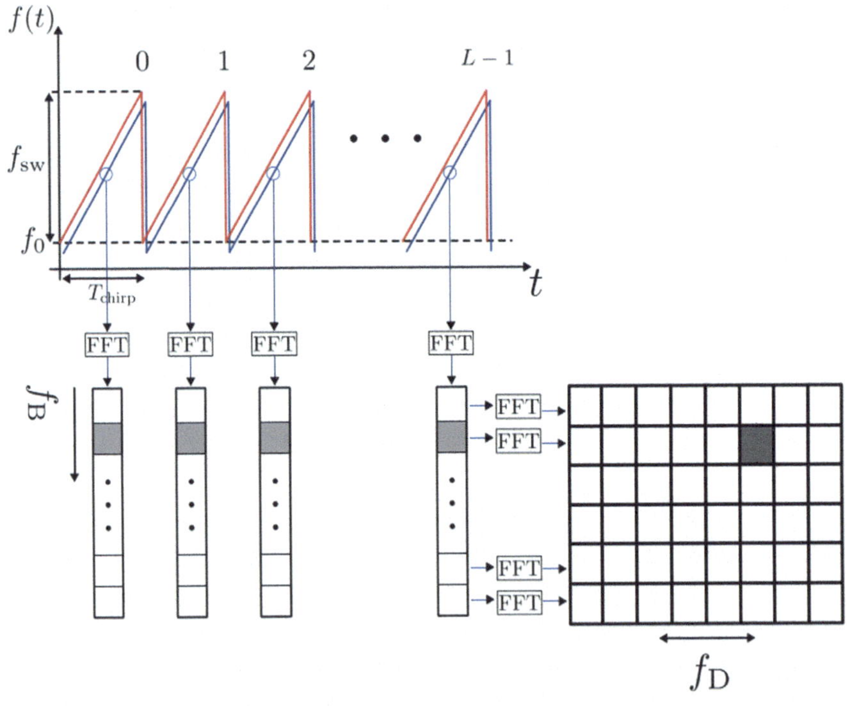


**Fig. 2.11** L number of same series of FM for FFT

### 2.4.4 MIMO System Millimeter Wave Radar

Multiple-Input Multiple-Output (MIMO) technology refers to the transmitting and receiving of signals through the use of multiple transmit and receive antennas where a signal could be transmitted or received via multiple antennas.

MIMO radar has M number of transmitters and N number of receivers. Base on the cross-sectional statistical characteristics, it can be considered constant when receiving multiple array elements, and hence this improves the system's performance in estimating the target parameters. Since MIMO radar can simultaneously generate or transmit Mth orthogonal transmission waveform and receive or process N number of received signals, by making use of the waveform diversity technique to multiply the virtual arrays, it is equivalent to increasing the aperture of the receiver. In comparison to traditional radar, MIMO radar could achieve higher angular resolution and spatial resolution. However, the MIMO radar system also requires a larger number of transmitters and receivers, hence, the practical application cost is high.

The MIMO array depicted in Fig. 2.12 consists of four equally spaced transmit antennas. The virtual array obtained by pairing the transmit and receive antenna positions is a uniform linear array with elements located at (Fig. 2.13):

$$x_k = kd, k = 0, 1, \ldots, NM - 1 \tag{2.29}$$

1. **Time Division Multiplexing (TDM)**

MIMO radar system adopts the working mechanism of the TDM. The switching between the transceiver and the transmitting and receiving array antenna unit, through the high-speed electronic switch, can reduce the number and complexity of the MIMO radar antenna, and hence, reduce the cost and obtain better direction of arrival (DOA) and the angular resolution or lateral resolution than conventional radar.

Consider a MIMO radar that uses the TDM operation mode with M number of transmitting antennas and N number of receiving antennas, the received signal at each receive antenna is divided into M number of virtual signals. When the number of reception antennas is N, the total number of dummy signals is MN and each virtual signal corresponds to a particular transmitter, m, and receiver, n, in a

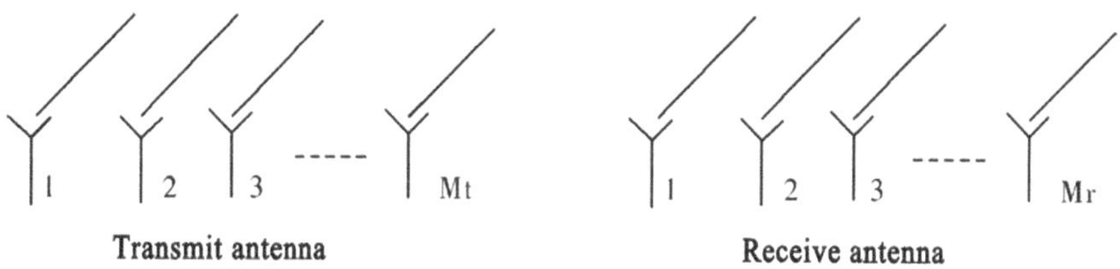


**Fig. 2.12** MIMO radar antenna array

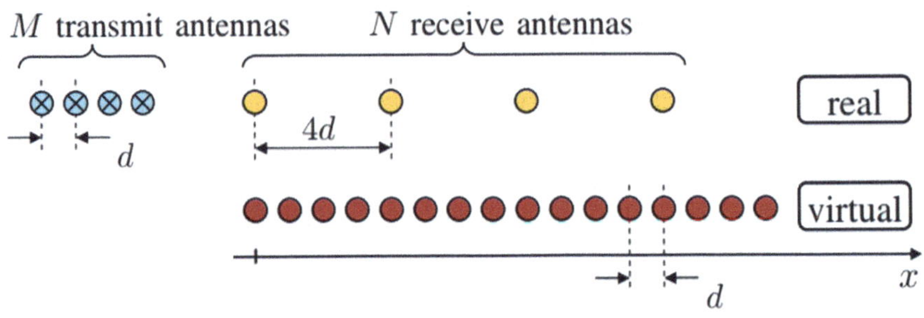


**Fig. 2.13** Virtual array diagram

matched pair, independent processing is conducted in terms of a single antenna system. As shown in Fig. 2.14, each transmit antenna could occupy a radar channel exclusively within a very short time frame. Different colors represent the duration of each time slot, $T_{slot}$, of the different transmit antennas. At each receive antenna, the received signal is sampled at a rate of $1/T_{slot}$. These signals have the same TDM structure as the transmitted signal and can directly identify the corresponding transmit antenna.

The signal processing structure eventually obtained a two-dimensional distance—Doppler matrix, $G_k(R,f_D)$, in a complex domain, where $k = 0, 1, \ldots, MN - 1$. Figure 2.15 illustrates this processing step.

Due to the spatial distribution of the antenna, the MN matrices obtained are same in size but of different phases:

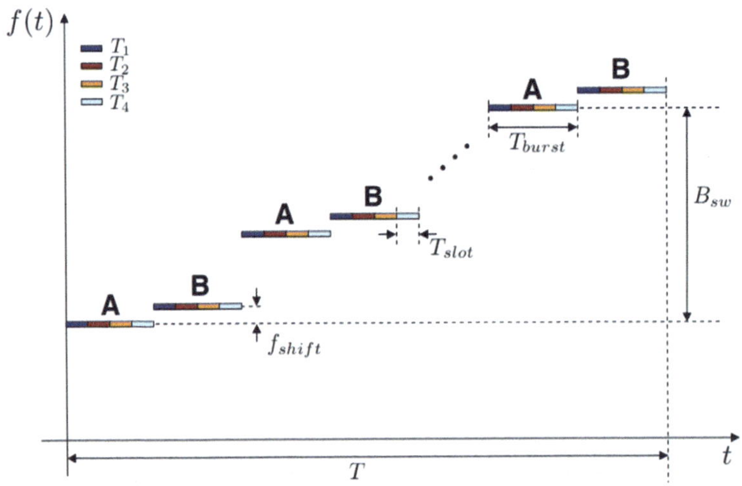


**Fig. 2.14** A radar signal waveform in time division multiplex mode

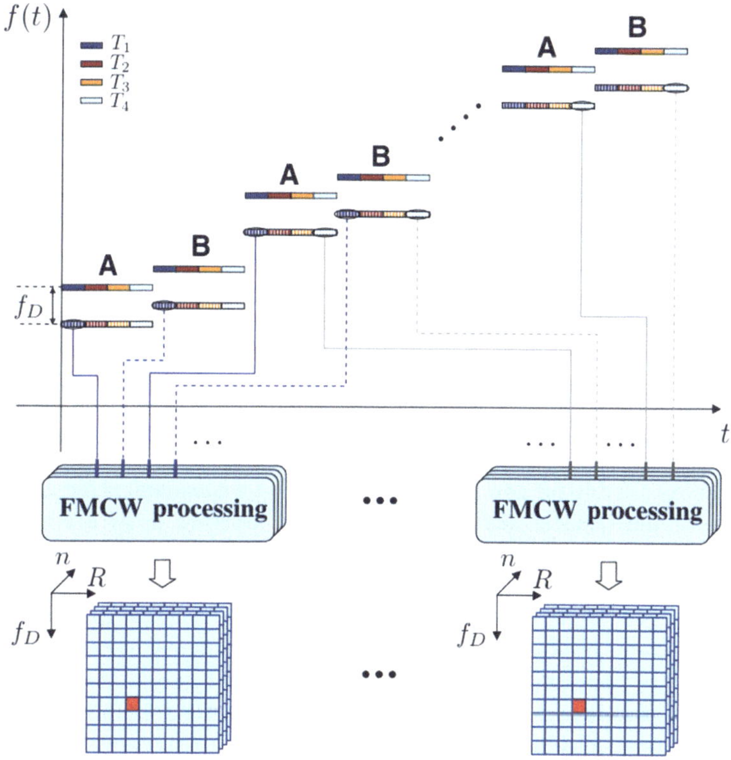


**Fig. 2.15** When M = 4, N = 4, the signal processing result of each virtual signal is the distance-doppler matrix

$$\arg(G_k(R,f_D))|_{R,f_D-const} = 2\pi\frac{x_k}{\lambda}\sin\theta + \phi_0 \tag{2.30}$$

Wherein $x_k$ is the kth coordinate of the virtual element, $\phi_0$ is a specific reference phase and is equal for all matrices. This result is the basis of a beamforming process with high azimuth resolution. It can be implemented by applying spectrum analysis on the matrix, $G_k(R,f_D)$, as shown in Fig. 2.16. The result is a three-dimensional matrix, $G_k(R,f_D,\theta)$, that also includes target range, radial velocity and azimuth information for multiple targets.

2. **Beamforming**

Beamforming is a classic array signal processing technique. It can be regarded as linear spatial filtering. Having a Nth antenna array of antenna elements, the array receives a signal from a fixed direction. Since the geometric positions of the antenna elements in space are different, the time at which the radio frequency signals reach each element is also different, which results in a phase difference between different received signals. However, if the complex baseband signal is assumed to be a

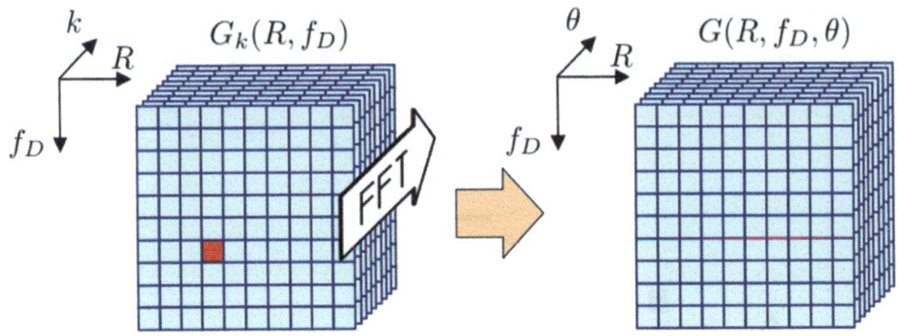


**Fig. 2.16** Beamforming process

narrowband signal, the baseband signal can be considered constant under such a short time difference. If the direction of the incident signal is known, the phase shift of the RF signal can be compensated by a phase shift or delay unit before summing the signal. As a result, the overall pattern will exhibit a maximum phase array in the direction of the incident signal. This method is referred to as a conventional beamforming.

If the complex baseband signal is not a narrowband signal, the baseband signal may also change during the time interval of the relative delay of receiving the RF signal. Therefore, each antenna element receives a different complex baseband signal. In this case, the technique that combines the use of the narrowband beamforming and time-domain filtered broadband beamforming must be adopted.

Beamforming techniques may also be used to reduce delay spread due to multipath propagation of a signal. To achieve this, the transmitting or receiving end uses a filtering forming technique to adjust its beam pattern such that there is no beam in the direction of the main propagation path of the remote reflector. Thereby, signal components having excessive delay in the received signal can be eliminated.

MIMO systems can be used simultaneously at both ends of the transceiver to receive and transmit the beam formed. Figure 2.17 illustrated the schematic diagram of a point-to-point MIMO beam forming transceiver system. In the MIMO system, in accordance with the beamforming implementation, it can be divided into an analog beamforming (ABF) and digital beamforming (DBF). The ABF is achieved through analog phase shifters in the radio frequency domain while the DBF is achieved through digital signal processing technique in the baseband. The beamforming weight vectors acquisition mode can be divided into the codebook beamforming (beam switching) and the adaptive beamforming.

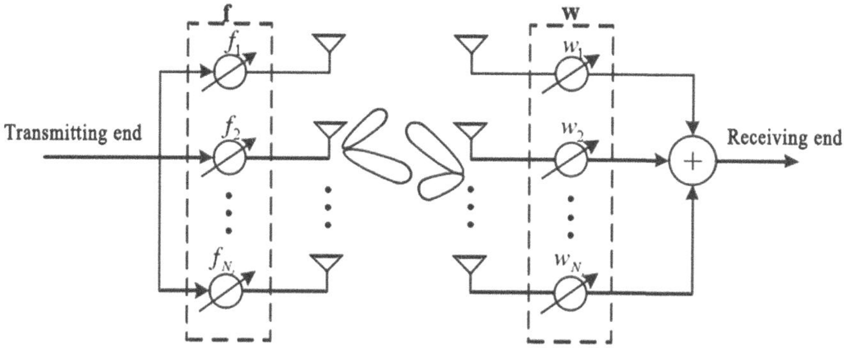


**Fig. 2.17** Schematic diagram of point-to-point MIMO beamforming

## 2.5 Related Technologies of the Millimeter-Wave Radar

### *2.5.1 Millimeter-Wave Radar Signal Processing Technology*

Radar signal processing technology is one of the key technologies of radar systems which realizes target detection and position estimation. The processing environment and target characteristics experienced by the radar sensors are different from those of traditional military radars. The radar detection environment is complex where there exists interference from trees and buildings. The target RCS is relatively large and the motion characteristics are relatively regular. Based on these characteristics, it is necessary to research the radar signal processing technology in a targeted manner. Firstly, it can improve the radar's ability to detect targets in the clutter background. Secondly, it can improve the radar's estimation accuracy of the target position. The following two key technologies for radar signal processing are introduced, namely, constant false alarm detector and MTD coherent accumulation processing.

1. **Constant False Alarm Detector**

The received signal received by the radar includes not only the target signal but also various noise, clutter and interference signals. When using a fixed threshold for detection, if the threshold is set high, it results in the false alarm rate to be low, and hence, a large number of missed alarms may occur. On the other hand, when the threshold is set low, although the probability of detection increases, many false alarms could arise due to noise, clutter and interference. In an unmanned system, in general, the target that the radar needs to detect is always affected by interference caused by ground objects, rain and snow, wave clutter and internal noise of the receiver. While detecting the presence of the target, if the threshold is fixed under an unstable cluttered environment, the probability of the false alarm will increase when the average power of the background clutter increases. This results in the saturation of the computer processing capability, and hence, affects the normal

**Fig. 2.18** Unmanned vehicle in snow environment

operation of the radar system. For instance, in the case of an unmanned vehicle system, as illustrated in Fig. 2.18, the false alarm rate is increased due to the different reflectance and roughness of the snow surface from the normal road surface that resulted in the snow clutter interference.

Therefore, in the modern radar signal processing, in order to improve the performance of the radar, it is first necessary to improve the signal-to-noise ratio and the signal-to-interference ratio at the input of the detector. The measures are to reduce the value of the noise at the receiver, and adopt various measures to suppress clutter and anti-interference. However, despite the above method, there still exists noise, clutter and interference components at the detector input. Since the internal noise level of the receiver is slowly changing due to the influence of the analog device, the clutter and interference residual are also time-varying and non-uniformly distributed in space, so various constant false alarm methods are still needed to ensure that the radar signal detection retains the constant false alarm characteristics (CFAR). The constant false alarm method uses an adaptive threshold instead of a fixed threshold. This adaptive threshold can be adjusted according to the background noise, clutter and interference. If the background noise, clutter and interference are large, the adaptive threshold is increased; and if the background noise, clutter and interference are small, the adaptive threshold is lowered to ensure that the false alarm probability is constant.

The internal noise of the receiver belongs to Gaussian white noise. After passing through the envelope detector, the noise voltage obeys the Rayleigh distribution and its probability density function is given as:

$$P(x) = \frac{x}{b^2}\exp(-\frac{x^2}{2b^2}) \tag{2.31}$$

In the formula, $b$ is the Rayleigh coefficient. The size of $b$ is proportional to the mean, $\mu$, of the noise $x$, and is given as:

$$b = \sqrt{\frac{2}{\pi}}\mu \tag{2.32}$$

As seen from the above equation, $P(x)$ is a function of $b$, and $b$ is related to the $\mu$ of the strength of the represented noise. Hence, if the signal is detected with a fixed threshold in the white noise background, the false alarm probability will change with the strength of the noise as shown in Fig. 2.19.

If $y = x/b$ is used to substitute $x$ in the formula, the noise voltage, $x$, will be normalized, then the probability density function y is given as:

$$P(y) = y\exp(-\frac{y^2}{2}) \tag{2.33}$$

As $P(y)$ is not related to the level of noise, hence, even when the fixed noise threshold is adopted for detecting the background signal, the false alarm probability will not change with the intensity of the input noise. As a result, a constant false alarm effect could be achieved. Based on this principle, a constant false alarm detector under the white noise background could be achieved as shown in Fig. 2.19.

In Fig. 2.19, a sampling pulse should be made to calculate the mean estimate of the data from the sample data in the rest period of the radar because the data is representative of the noise and generally do not contain the target and clutter signals. Additionally, the calculation of the mean estimation, $\hat{\mu}$, requires a large number of noise data samples, and the number of noise data samples in a single radar rest period is limited, so the rest period data samples of multiple radar repetition periods are often used for calculation. To simplify the calculation, the mean estimation result between adjacent repetition periods can be smoothed by a first-order recursive filter, as shown in Fig. 2.20.

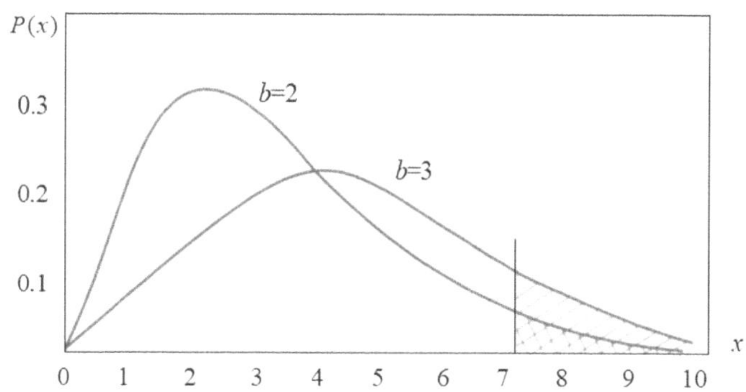


**Fig. 2.19** Fixed threshold under white noise background

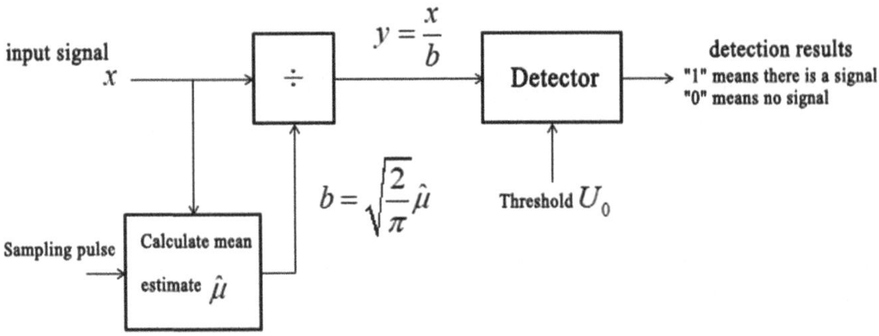


**Fig. 2.20** Constant false alarm detector under white noise background

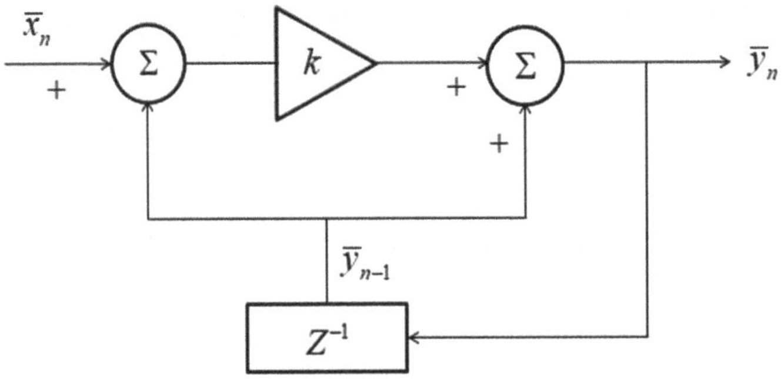


**Fig. 2.21** First-order recursive filter

In Fig. 2.21, $Z^{-1}$ represents the cross-firing cycle delay, $\bar{x}_n$ represents the mean of all the radar data samples in the nth repetition rest period, and $\bar{y}_n$ represents the output of the recursive filter, which is given as:

$$\bar{y}_n = K(\bar{x}_n - \bar{y}_{n-1}) + \bar{y}_{n-1} = (1 - K)\bar{y}_{n-1} + K\bar{x}_{n-1} \tag{2.34}$$

Figure 2.20 shows that for the CFAR detector with the white noise background, it calculates the mean estimation of the input noise and then proceed with detection by normalizing the input signal, $x$. At this point in time, the detection threshold can be a fixed threshold. In order to avoid the input signal, $x$, normalization, the detection threshold, $U_0$, could be adapted according to the magnitude of the mean estimation, $\hat{\mu}$, to obtain a CFAR detection effect.

The recursive filter outputs a mean estimation of the noise, $\hat{\mu}$, from the input signal, $x$, which is then multiplied by the threshold multiplier, $K$, to obtain an adaptive threshold of $U_0 = K\hat{\mu}$. The threshold multiplier, $K$, is a scalar and the size should be determined according to the size of the false alarm probability as required.

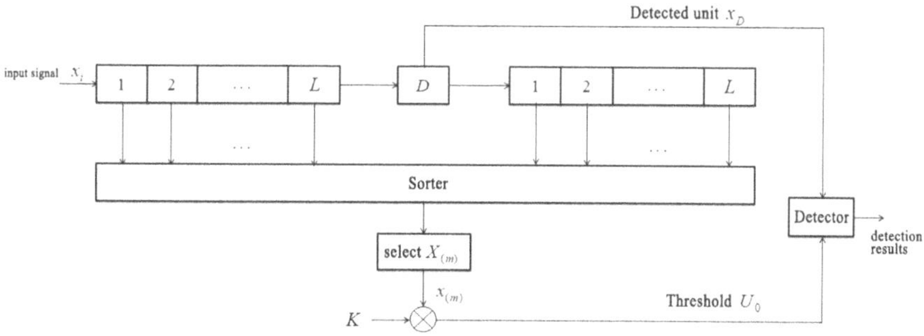


**Fig. 2.22** Configuration of the ordered constant false alarm detector

If other target signals (interference targets) appear in the reference unit, it will cause a decrease in the detection performance of the constant false alarm detector. In order to increase the anti-interference ability of the CFAR detector, the ordered statistic constant false alarm rate is proposed and the configuration is shown in Fig. 2.22.

The sequencer within a reference cell performs sequencing of $2L$ number of $x$ value according to magnitude. The data sequence after data sorting is given as:

$$x_{(1)} \leq x_{(2)} \leq \cdots \leq x_{(2L)} \tag{2.35}$$

After sorting, select the mth sample, $x_{(m)}$, as the clutter estimation within $2L$ number of the reference units. The threshold, $U_0$, is obtained after $x_{(m)}$ multiplies with the threshold multiplier, $K$, i.e. $U_0 = Kx_{(m)}$. Under general circumstance, $m$ could be taken as $^3/_3$ of the reference unit value, $2L$, and hence the formula is given as:

$$m = \frac{3}{4} \times 2L = 1.5L \tag{2.36}$$

When there is one or more interference targets entering $2L$ number of reference units, it will only change the sorting result from the OS-CFAR detector, however, the impact on the threshold is small.

Under the Rayleigh clutter condition, after going through the square-law detector, $x_i$ will follow the exponential distribution and an ordered CFAR detector false alarm probability, $P_f$, and the relationship between the reference units $2L$ and $m$ could be obtained:

$$P_f = mC_{2L}^{m} \frac{\Gamma[2L - K + 1 + K][m]}{\Gamma[2L + K + 1]} \tag{2.37}$$

From the Eq. (2.37), $C_{2L}^{m}$ represents the selection of $m$ number of combinations from $2L$ number of reference units, where $C_{2L}^{m} = 2L!/(2L-m)!m!$ is a gamma function. In the formula, the false alarm probability, $P_f$ is independent of the clutter power, and hence, the constant false alarm effect could be obtained.

2. **MTD Coherent Accumulation Processing**

In the unmanned systems, other than the clutter and noise that could affect the radar's false alarm rate to increase, the radar could also be affected by various clutter. For instance, the ground clutter caused by various stationary non-target reflectors on the ground, because most of them are stationary reflectors, their Doppler frequency is near zero frequency. For moving targets, the speed of motion will cause a certain Doppler shift. For multiple pulse echoes, the echo phase of the moving target is coherent between multiple pulses. Therefore, if the echo phase of the moving target can be compensated, the multiple pulses can be coherently added to obtain the largest energy accumulation. As the ground echo does not have the above characteristics, hence, it cannot be effectively accumulated.

The use of coherent accumulation to improve the detection performance of moving targets in the clutter background is achieved by using the difference between the spectrum of the target echo signal and the spectrum of the ground clutter. The former has a certain Doppler frequency, while the Doppler frequency of the latter is zero. The two are separated from each other in the spectrum. Therefore, for non-coherent accumulation, since it only utilizes the echo characteristics of the target, it is impossible to effectively distinguish the target and ground clutter from the frequency domain. In the unmanned vehicle system as shown in Fig. 2.23, the dynamic target detection capability of the radar is highly demanded and is carried

**Fig. 2.23** Unmanned vehicle dynamic target scene

out in the context of ground clutter interference caused by a large number of stationary reflectors. The coherent accumulation process can effectively improve the radar target detection ability under the clutter background.

Commonly used coherent integration processing methods include the Moving Target Indicator (MTI) and the Moving Target Detection (MTD). MTI processing, also known as multi-pulse cancellation processing, mainly uses a number of pulse repetition periods of data for clutter suppression processing. The MTD processing uses the Doppler filter bank to perform the in-phase accumulation processing of the moving target, and at the same time, suppresses the ground clutter by using the difference between the Doppler velocity of the moving target and the Doppler velocity of the ground clutter. MTD processing can obtain higher signal-to-noise, ratio improvement factor, and signal-to-noise ratio improvement factor. Hence, MTD processing is mainly discussed here.

Filter bank, which is MTD core processing, determines the MTD processing performance. According to different implementation methods, the MTD filter bank can be divided into the FFT method and the FIR method. The FFT method is simple to implement and has higher execution efficiency, but the design flexibility is relatively poor. In order to suppress the high sidelobe level, the windowing processing is usually required. The FIR method is designed with high flexibility, and filter banks with special frequency characteristics can be selected according to the design requirements.

After the main beam of the radar illuminates the ground area, the echo signal reflected from the ground is called the ground clutter. The strength of the ground clutter is related to the transmitter power, the gain of the main beam of the antenna, the reflection ability of the ground object, the height of the antenna and other factors. The intensity may be several tens of dB higher than the noise of the radar receiver.

Usually the echo signal from the ground clutter is a random process. Its power spectrum can be approximated as:

$$C(f) = G_0 \exp\left(-\frac{f^2}{2\sigma_f^2}\right). \tag{2.38}$$

where $G_0$ is a constant which determines the intensity of the clutter spectrum; and f is the standard deviation of the clutter power spectrum which determines the width of the clutter spectrum.

The clutter spectrum approximates a Gaussian spectrum and is distributed around the zero frequency. The frequency in the above formula represents the analog frequency which is the actual frequency of the signal. Generally, radar signal processing focuses on digital signals and the MTD processing is performed between pulses. For discussion convenience, the frequency is normalized by the pulse repetition frequency (PRF) of the radar.

The spectrum of the ground clutter is located near the zero frequency, and the frequency domain of the moving target is distributed near its corresponding

Doppler frequency. Therefore, the difference between the two in the frequency distribution can effectively distinguish the moving target from the ground clutter. To this end, a set of narrow-band filter banks that are adjacent and overlapping in frequency are often used in engineering to realize the resolution of moving targets, which is known as MTD processing.

The implementation of the MTD filter bank is divided into the FFT method and the FIR method. For the FFT method, it uses multiple frequency channels that are formed by the FFT transform; the FIR method is to design multiple band pass filters that have a specified center frequency. In fact, multiple frequency channels formed during FFT transformation can also be considered as a set of FIR filter banks. MTD implementation structure could be illustrated as shown in Fig. 2.24 (assuming the MTD process has M number filter).

For a given MTD filter bank, the indicators that are commonly used to examine its performance include signal-to-noise ratio gain (improvement factor), signal-to-noise ratio gain, and so on. In fact, when the clutter spectrum model and MTD parameters are given, the performance index of the MTD can be calculated.

Assume that the data input to the MTD filter bank is expressed as:

$$x(l) = x_s(l) + x_c(l) + x_n(l) \tag{2.39}$$

where $x_s(l)$ represents signal, $x_c(l)$ represents clutter and $x_n(l)$ represents noise.

In order to examine the processing performance of the MTD, it is usually assumed that the signal is a continuous wave sinusoidal signal, and could be expressed as:

$$x_s(l) = A_s \exp(j2\pi f_d l) \tag{2.40}$$

where $f_d$ (normalized frequency) is the center frequency of the signal.

Additionally, the power spectrum for clutter is assumed as shown in Eq. (2.40); and complex Gaussian white noise is assumed.

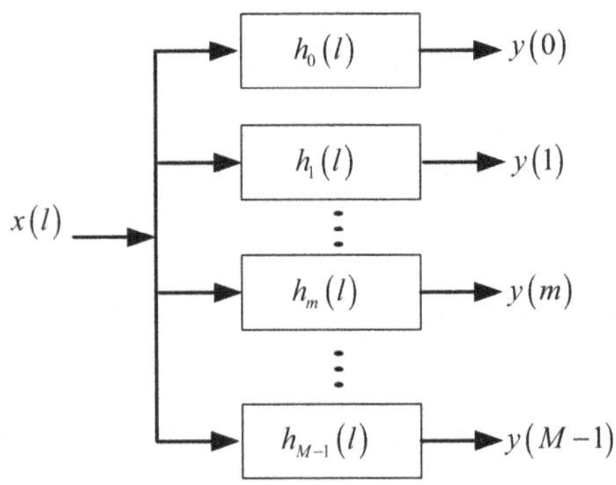


**Fig. 2.24** MTD implementation structure

Assuming that the first mth MTD filters (also known as channels) coefficient is $H_m(L)$ (where L = 0, 1,..., 1 − N; and N is the order of the filter), then the first m number of output channels can be expressed as:

$$\begin{aligned} y(k) &= y_s(l) + y_c(l) + y_n \\ &= \sum_{l=0}^{N-1} h_m(l)x_s(l) + \sum_{l=0}^{N-1} h_m(l)x_c(l) + \sum_{l=0}^{N-1} h_m(l)x_n(l) \end{aligned} \tag{2.41}$$

Based on the above assumptions, and by using Eq. (2.41), the improvement factor for the mth MTD channel could be obtained.

$$G_{sc}(f_d, m) = \frac{\sum_{l=0}^{N-1} \sum_{p=0}^{N-1} h_m(l)h_m^*(p)\exp[j2\pi f_d(l-p)]}{\sum_{l=0}^{N-1} \sum_{p=0}^{N-1} h_m(l)h_m^*(p)\exp\left\{-2\sigma_f^2[\pi(l-p)]^2\right\}\exp[j2\pi f_0(l-p)]} \tag{2.42}$$

The signal-to-noise ration could then be expressed as:

$$G_{sn}(f_d, m) = \frac{\sum_{l=0}^{N-1} \sum_{p=0}^{N-1} h_m(l)h_m^*(p)\exp[j2\pi f_d(l-p)]}{\sum_{l=0}^{N-1} |h_m(l)|^2} \tag{2.43}$$

In the actual radar system, parameters such as the number of channels and the MTD filter coefficients are predetermined, but the Doppler frequency (motion speed) of the target is unknown. It can be seen from the formula that the corresponding MTD processing performance is different for targets with different Doppler frequencies. Therefore, it is necessary to examine the performance of the MTD processing over the entire pulse repetition frequency.

The following method to define the frequency response curve of the MTD filter bank is used:

(1) Sequentially generate targets with different normalized Doppler frequency;
(2) Determine the channel where the target outputs the maximum energy among N number of MTD channels;
(3) Compute the power of the signal, clutter and noise that is outputted by the channel to calculate the gain, and thereby, obtain the performance parameters that correspond to a certain Doppler frequency;
(4) Draw the performance parameters for all the Doppler frequencies to obtain the Doppler frequency response curve of the MTD;
(5) Average each of the performance parameters of all frequencies to obtain the mean of the parameters (i.e. signal-to-noise ratio gain and signal-to-noise ratio gain).

The channel number with the highest signal output energy is selected becauseusually after the MTD processing, the constant false alarm processing of the frequency channel will be used and the Doppler frequency of the target always falls on multiple frequencies as the MTD filters overlap each other. On the other hand, the ground clutter is always near the zero frequency. Therefore, the channel with the largest output signal energy is selected for constant false alarm processing with the highest signal-to-noise (noise) ratio.

## *2.5.2 Millimeter-Wave Radar Data Processing Techniques*

Millimeter-wave radar data processing and radar signal processing are important components of modern radar systems. Signal processing is used to detect targets, using a certain method to obtain various useful information of the target, such as distance, speed and shape of the target. The data processing can further process the target's points and tracks, and predict the position of the target in the future to form a reliable target trajectory, so as to achieve real-time tracking of the target.

### 2.5.2.1 Millimeter-Wave Radar Point Data Processing

The points obtained by the millimeter-wave radar are multi-valued in both distance and azimuth. Therefore, after the radar is detected, the same target usually has multiple original point data in the distance and the azimuth. In order to achieve multi-batch and multi-target positioning detection by the radar and to ensure track data processing is accurate and reliable, there is a need to conduct the points condensing process for the point tracks. Since the obtained original measurement of the target is not unique, it is necessary to perform the merging and resolution processing on the distance and the azimuth before the dot condensing process, so that one target corresponds to s individual measurement (Fig. 2.25).

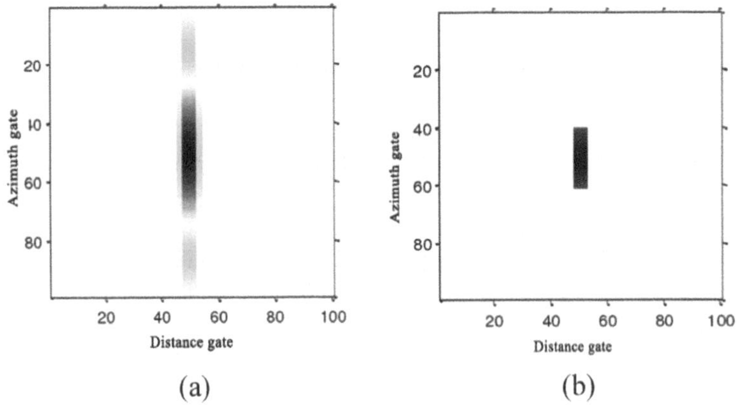


**Fig. 2.25** **a** Radar original trace data; **b** original trace merge and resolution processing

After merging and resolution processing of the original trace points, data in all directions are coagulated to obtain a unique value in distance. Thereafter, to perform azimuth coagulation process on point track data to obtain only one orientation estimate. Finally, the target unique estimate is calculated by the interpolation formula. Assume that the azimuth unique value is obtained between the i and i + 1 point, the interpolation formula is then given as (Fig. 2.26):

$$R_0' = R_i + (R_{i+1} - R_i)\frac{\theta_0 - \theta_i}{\theta_{i+1} - \theta_i} \tag{2.44}$$

#### 2.5.2.2 Millimeter-Wave Radar Track Data Processing

Millimeter-wave radar track data processing begins by finding the motion trajectory of the target from the points detected by the radar at every moment so as to track or predict the motion of the target. The millimeter-wave radar track processing includes the following: the initiation of the radar tracking, radar track data correlation and radar track filtering. The initiation of the millimeter-wave radar tracking is to establish the track of the target motion from a large number of points. The millimeter-wave radar track data correlation determines the returned signal of the highest likelihood to become the target point from a multiple of returned signal candidates. The millimeter-wave radar tracking filter estimates the position of the next target point based on the existing track.

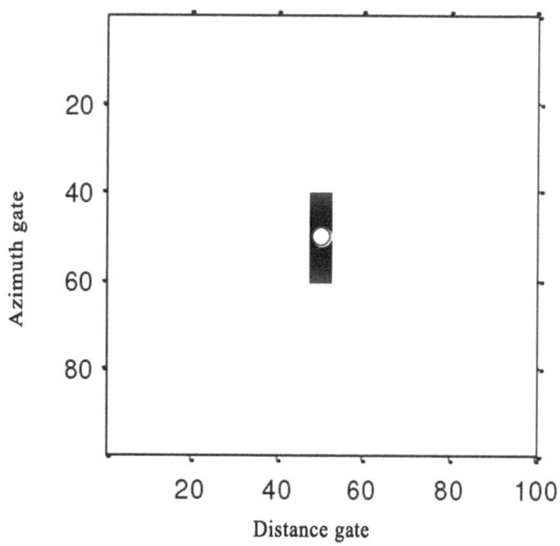


**Fig. 2.26** Target point unique estimate

1. **Track Initiation**

The millimeter-wave radar track initiation is the process in which the target enters the radar monitoring range (being detected) to establish a track. Track initiation is an important issue in radar track processing. If the track initiation is not correct, there is no way to achieve the correct track. The main task of the track initiation is to establish the target track whether in a quiet or noisy environment. In practical applications, where the environment is more complicated and the number of possible tracks is too large, the most commonly used methods include the intuitive method and the logic method in order to ensure the speed and efficiency of the track initiation. Other track initiation algorithms, due to their high complexity or prior knowledge, are rarely used in practical engineering.

(1) Intuitive Method

The intuitive method uses the maximum speed and the minimum speed to establish the wave gate, which is simple and easy to implement, and is the most commonly used algorithm for track initiation. In adjacent scan cycles, the largest target acceleration constraints could be added between measurements. Assume that $r_i(i = 1, 2, \ldots, N)$ is obtained by continuously scanning N number of times, if there are more than M times of fulfilling the two conditions listed below within N times of continuous scanning, then the track initiation can be considered as successful. The two conditions are:

(1) Measure or estimate the speed value between the values of Vmin and Vmax, that is, the target falls into the ring-shaped gate formed by this speed constraint, as shown in Fig. 2.27. This determination is based on the first scan that is particularly suitable for the measurement, or a measurement that is not associated with all of the tracks obtained in subsequent processes.
(2) The acceleration measured in the ring gate is analyzed for its acceleration, which should be less than the maximum possible acceleration, $a_{max}$. If there are multiple returned signals, the returned signal with the least acceleration is selected as the effective measurement.

**Fig. 2.27** Ring gate

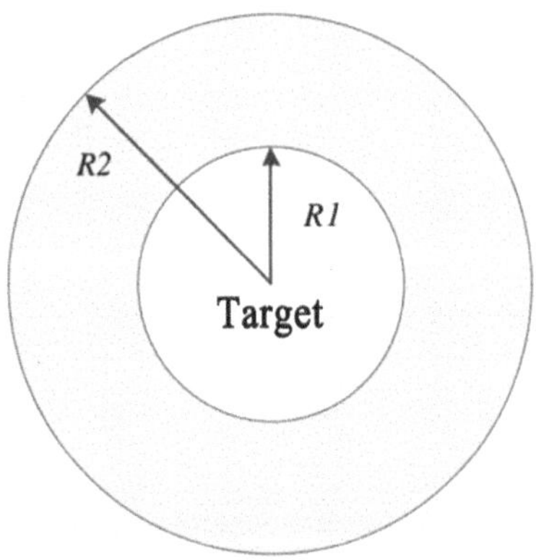

(2) Logic Method

The logic method is a track initiation method that can be applied throughout the track processing. There are some similarities between logic method and intuitive method where track initiation is considered to be successful if the number of correct detections is not less than a certain number of measurements in the Nth scanning period. If the number of detections that meet the threshold is not sufficient within a window period, the window is shifted back. The difference between the two methods lies in the wave gate, whereby the intuitive method uses velocity and acceleration as the gate limit at the start of the track and the logic method determines the existence of the track by predicting the future position of the target and setting the relevant gate limit (Fig. 2.28).

$$d_{ij}^{l}(t) = \max[0, z_j^{l}(k+1) - z_j^{l}(k) - v_{\max}^{l\max[t_{j_{\max}}^{l^l}} \tag{2.45}$$

2. **Track Data Correlation**

Track-track data correlation problem is the fundamental and core problem for radar track processing, especially when the target motion trajectories cross each other, multiple targets are densely distributed, the radar itself has large measurement error, the detection probability is less than 1, and there exists strong clutter with non-linear systems, the data correlation becomes more difficult.

Conducting data correlation under a cluttered environment, since the returned signal could come from the target, clutter or false alarm, could adversely affect the correlation between the returned signal point and the target point due to the clutter and measurement noise. Data correlation is based on the state of the returned signal and target state to find the returned signal of the next most possible location of the target.

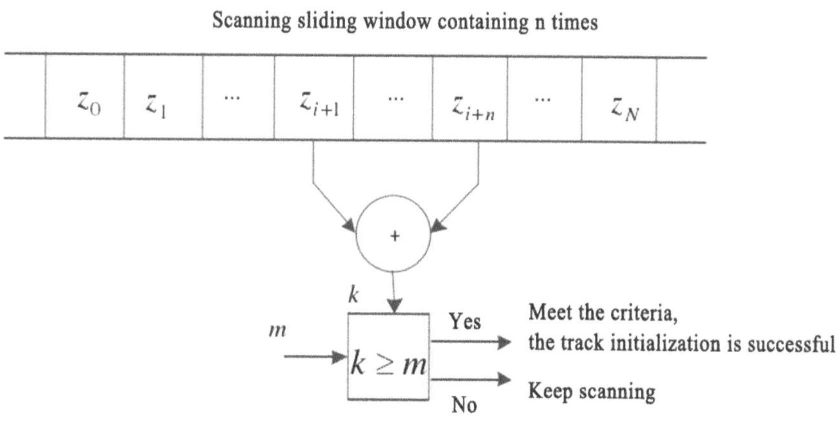


**Fig. 2.28** Schematic diagram of the logic method

(1) Nearest Neighbor Standard Filter

Nearest Neighbor Standard Filtering (NNSF) is a simple and effective algorithm. The algorithm uses the prior statistical characteristics to estimate the correlation performance. Firstly, the tracking gate (correlation gate) is set. The tracking gate is used for the measurement, and the measurement that falls in the tracking gate and is closest to the position of the track prediction point will be selected as the new track point. Track points are updated in real time. Assume $z_k = \{z_{k,i} : i = 1, 2, \ldots, m_k\}$, where $\widehat{z_k}$ is the track prediction point at time k, the tracking gate setting is then given as:

$$[z_k - \hat{z}_k]' S_k^{-1} [z_k - \hat{z}_k] \leq \gamma^2 \tag{2.46}$$

The problem with using the nearest neighbor standard filter is that only the measurement that falls closest to the predicted value is updated into the track. When there exists more clutter or noise, the track data caused by the clutter or noise is also updated which causes the tracking error or loss. Therefore, the nearest neighbor standard filter is more applicable when the radar works in a high signal-to-noise ratio with few targets environment.

(2) Probabilistic Nearest Neighbor Method

The Probabilistic Nearest Neighbor Method (PNNF) applies the idea of probability theory to the nearest neighbor correlation algorithm, and also considers the nearest neighbor measurement as the true measurement of the target. However, the algorithm considers that the nearest neighbor measurement could be derived from clutter and that there are no measuring points in the wave gate, and hence proceeds to adjust the state error covariance. Three conditions are illustrated as follow: (1) No measurement falls into the gate (M0); (2) the nearest neighbor in the relevant gate is derived from the target (MT); and (3) the nearest neighbor is derived from the false alarm.

3. **Track Tracking**

The tracking filter uses the observation value in the effective observation time, and obtains the state estimation value of the linear discrete-time system by selecting the appropriate estimation method. With the continuous acquisition of the observation value, the state estimation value of the system is continuously obtained and the system state is continuously being monitored to obtain the target's continuous track. The common tracking filters that are used for radar data processing include the least-squares filer, Kalman filter and the $\alpha - \beta$ filter.

(1) Least Squares Filter

Radar data processing could conduct filtering and prediction through the use of observation data and a function that integrates certain rules. Since there exists error in the observed data, it is unreasonable to pass all known points through the approximation function as it is equivalent to retaining all data errors. The principle of the least-squares filter is based on the following process: From a given data set, select the approximation function form and given the function class, H, $\varphi(\mathrm{x}) \in \mathrm{H}$, and the function is then given as:

$$\sum_{i=1}^{n}[y_i - \varphi(x_i)]^2 = \min_{\varphi \in H} \sum_{i=1}^{n}[y_i - \varphi(x_i)]^2 \tag{2.47}$$

This method of approximation function is called the data fitting least squares method, and the function, $\varphi(\mathrm{x})$, is called the least squares function of the dataset. Usually H is a collection of relatively simple functions, such as low-order polynomials and exponential functions. Polynomial fitting is generally used in radar data processing.

(2) Kalman filter

During the radar data processing, the observations, $z^k = \{z_0, z_1, \ldots, z_k\}$, that are obtained within the limited observation timeframe are used to estimate the state of the linear discrete time dynamic system, s. If the system model assumes that the state equation is satisfied, then:

$$s_{k+1} = \Phi_k \hat{s}_k + B_k u_k + G_k v_k \tag{2.48}$$

In Eq. (2.48), $s_k$ represents the system state of n-dimensional vectors in k time period; $\Phi_k$. is the $n \times n$ state transition matrix in k time period; $u_k$ is the p order input matrix; $B_k$ is the n × p order input matrix; $v_k$ is the q-dimensional random vector that satisfies the Gauss white noise distribution; and $G_k$ is the n × q dimensional real-valued matrix where

$$E\{v_k\} = 0 \tag{2.49}$$

$$E\left\{G_k v_k v_j^T G_j^T\right\} = Q_k \delta_{kj} \tag{2.50}$$

The observation equation is also a linear function, i.e.

$$z_k = H_k s_k + L_k \omega_k \tag{2.51}$$

In Eq. (2.51), $z_k$ is the m-dimensional observation vector at the time k; $H_k$ is the m × n order observation matrix; and $\omega_k$ is the m-dimensional measurement noise which satisfies the Gauss white noise distribution and is given as:

$$E\{\omega_k\} = 0 \tag{2.52}$$

$$E\left\{L_k \omega_k \omega_j^T L_j^T\right\} = R_k \delta_{kj} \tag{2.53}$$

Additionally, assume $v_k$ and $\omega_k$ to be independent and hence satisfies:

$$E\{v_k\omega_k^T\} = 0 \tag{2.54}$$

According to the minimum mean square error, the system state prediction equation, state equation of the filter, the filter gain equation, and the residual covariance matrix are usually refered to as the Kalman filter equations. The derivation of the Kalman filter equation can be based on the assumptions of the linear model and the Gaussian distribution, and the best estimation criterion can be obtained by applying the best estimation criterion. The linear mean square estimation can also be adopted without making any assumptions about the distribution function of the process (Fig. 2.29).

(3) $\alpha - \beta$ Filter

When the target model uses the constant velocity model which follows:

$$K_{k+1} = \hat{P}_{k+1/k}H_{k+1}^T\theta_{k+1}^{-1} = \hat{P}_{k+1/k}H_{k+1}^T(H_{k+1}\hat{P}_{k+1/k} + R_{k+1})^{-1} \tag{2.55}$$

The filter gain matrix and constant matrix could be expressed as $K = [\alpha\beta/T]^T$, which is also known as the $\alpha - \beta$ filter device. Based on the given process noise and observation noise, the Kalman filter equation could be used to obtain the relationship between the known parameter values of $\alpha$ and $\beta$.

**Fig. 2.29** Application of Kalman filter tracking unmanned boats

### 2.5.3 *Millimeter-Wave Radar Imaging Technology*

The current automotive millimeter-wave radar products are mainly 24 or 77 GHz narrow-band vehicle radars, which can only obtain the target distance, speed, and azimuth angle information but not accurately identify the target. For instance, it cannot accurately identify whether the target is a car or a street sign. The high-distance resolution relies on the use of ultra-wideband signals, and the speed resolution is improved by long-term pulse accumulation. The angular resolution is improved by means of radar multi-channel transceiver array antenna or angle super-resolution processing technology.

The multi-channel, ultra-wideband, high-resolution and high-precision radar uses the 79 GHz ultra-wideband frequency millimeter-wave for target information acquisition, enabling centimeter-range resolution range. The 79 GHz high-resolution and high-precision millimeter-wave radar uses the three-dimensional imaging technology to process the high-resolution target information around the vehicles for the imaging. Hence, obtaining point cloud information and visual image information of the precise position and velocity of the target environment of the vehicle. At the same time, the three-dimensional image can be further integrated with the point cloud data of the lidar and the image information of the camera. Hence, the multi-channel, ultra-broadband, high-resolution and high-precision three-dimensional imaging millimeter-wave radar can greatly improve the intelligent vehicle's situational awareness effectiveness and stability.

In the real-time imaging technology of the millimeter-wave radar, the main technologies that are used include the mechanical scanning technology, synthetic aperture radar (SAR) technology, inverse synthetic aperture radar (ISAR) technology, digital beamforming (DBF) technology, multi-channel multiple-input-multiple-output (MIMO) technology, Orthogonal Frequency Division Multiplexing (OFDM) technology, etc. The following sections provide a brief introduction to Synthetic Aperture Radar (SAR) technology, Digital Beamforming (DBF) technology, and Multi-Channel Multiple-Input-Multiple-Output (MIMO) technology.

1. **Synthetic Aperture Radar (SAR) technology**

Synthetic Aperture Radar uses a small antenna to move at a constant speed along the trajectory of the long line array and radiate the coherent signal, and coherently process the returned signals that are received at different positions to obtain a higher resolution imaging. Due to the motion of the radar, different scattering points of the object have different radial velocity components, which can increase the overall resolution of the radar by increasing the speed resolution through long time pulse accumulation. Synthetic Aperture Radar requires the object to be stationary relative to the radar and for the radar to move at a uniform speed. Therefore, Synthetic Aperture Radar is mainly used for airborne or spaceborne radar imaging. In 2004, French R. Giret et al. realized a three-dimensional model for drones which uses a linear array of receiving antennas in conjunction with the syntactic aperture radar (SAR) technique and the chirp (LFM) pulse waveform.

In 2012, Markus Andres et al. from the University of Ulm, Germany, realized a three-dimensional imaging radar automobile for detecting a scattering centrers using mechanical scanning antenna array as shown in Fig. 2.30. They combined the synthetic aperture radar (SAR) technique and the digital beamforming (DBF) technique for three-dimensional imaging on different automobile. 2 GHz bandwidth radar can perform a 10 m distance imaging of a vehicle using a 2° angular resolution (Fig. 2.31).

2. **Beamforming Technique for Multi-Channel Transceiver Antenna**

The multi-channel, ultra-wideband, high-resolution and high-precision imaging radar uses a radar with a high frequency, a large bandwidth, and a high modulation difficulty. Therefore, a stable high-speed RF circuit and a signal processing circuit are required. At the same time, because high-resolution and high-precision imaging radar needs to process multi-channel sampling rate signals in real-time, and hence, large point cloud data and target information are generated. Therefore, the radar signal processing algorithm is currently one of the most popular and most challenging research area in the development of the millimeter-wave radar.

Advances in signal digitization and fast digital signal processing techniques that had resulted in cost reduction in the recent years have made digital beamforming (DBF) an increasingly important radar imaging technology. Digital beamforming (DBF) technology is a method to improve the resolution of the radar angles. The technology of electronic beam scanning which uses the multi-channel transceiver array of antennas has gradually become mainstream because it does not require the mechanical beam scanning. The University of Karlsruhe, Germany, introduced a radar system for three-dimensional object detection and imaging, and proposed a three-dimensional imaging principle based on two-dimensional digital beamforming (DBF). The three dimensional imaging radar is implemented by the combination of the frequency-modulated continuous wave (FMCW) distance ranging and the digital input beamforming (DBF) multiple-input-multiple-output (MIMO)

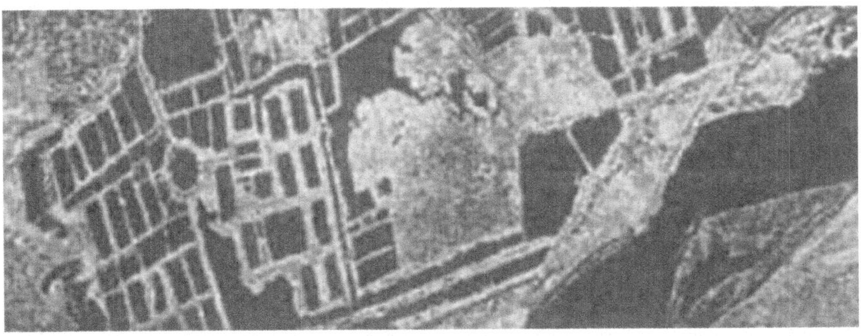

**Fig. 2.30** Composite image of airborne SAR technology (yellow river fishery)

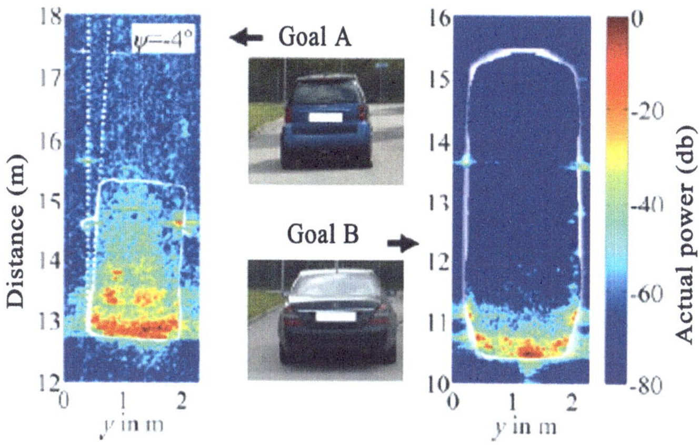


**Fig. 2.31** Three-dimensional imaging radar using SAR technology where the left picture is the target A imaging, and the right picture is the target B imaging

technology. The transmitting and receiving antenna arrays are arranged in quadrature, and are formed by the digital beamforming at the transmitter and receiver to obtain the two-dimensional azimuth and elevation angles. Through digital signal processing, digital beamforming (DBF) offers the possibility of simultaneously using multiple channels to collect the radar response signals and at the same time focuses on different angles (Fig. 2.32).

Currently, the 79 GHz ultra-wideband millimeter-wave imaging technology is still in research exploration stage. A more feasible technical solution is to fully utilize the ultrawide bandwidth and the internal RF chip that is integrated with

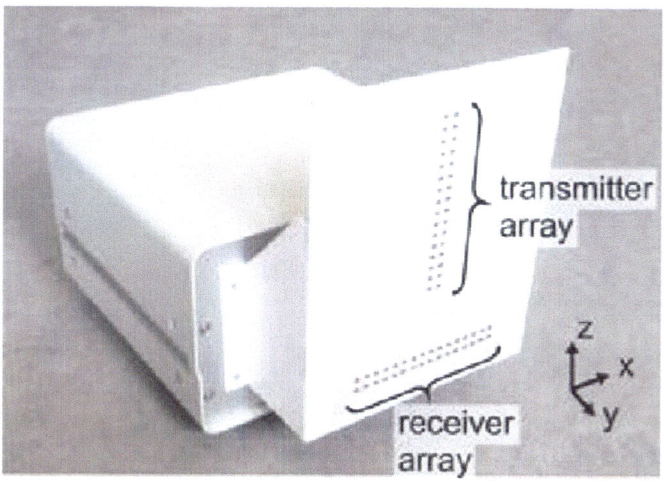


**Fig. 2.32** Imaging radar from Karlsruhe Institute of Technology, Germany

high-speed sampling analog-digital converters in the 79 GHz millimeter-wave radar, through the use of the high-speed compression of the ultra-wideband pulse technology, to achieve high range and high-speed resolution. By analyzing the distance ranging methodology for 79 GHz signal synthesis broadband and using the key technology of millimeter-wave bandwidth extrapolation (BWE), the all-pole signal model which is suitable for millimeter-wave automotive radar for road surface targets is obtained. The model parameters are adjusted in the sub-bands. The sub-bands with delay and phase difference are coherent. The coherent sub-band signals are interpolated and the standard pulse compression method is used to obtain the super-resolution distance image, thereby realizing high-resolution and high-precision centimeter-level ranging. The process is illustrated as shown in Fig. 2.33.

3. **Multi-Channel Multiple-Input-Multiple-Output (MIMO) Technology**

Multi-channel multiple-input multiple-output (MIMO) technology achieves high-resolution imaging by increasing the radar angular resolution through the use of multiple antennas at the transmitting and receiving ends. In 2010, the University of Connecticut proposed a concept based on the multi-channel multiple-input multiple-output (MIMO) Orthogonal Frequency Division Multiplexing

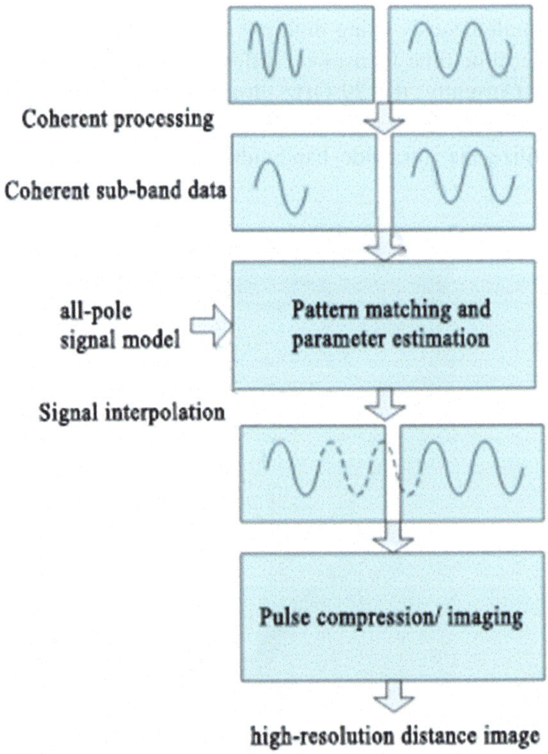


**Fig. 2.33** High range resolution processing workflow

**Fig. 2.34** Imaging of millimeter-wave radar imaging of a road scene

(OFDM-based) three-dimensional radar. Transmitted Signal Orthogonal Frequency Division Multiplexing (OFDM) signals have the feature that multiple transmitters radiate only at their own unique subcarriers while preserving the entire bandwidth and range resolution. The antennas are arranged in a "T" shape, and beamforming techniques and new radar processing techniques are used at the receiver to estimate the distance, azimuth and elevation angle information of the target by simultaneous transmission. Since the Orthogonal Frequency Division Multiplexing (OFDM) signals are orthogonal, the spectrum division strategy can be simultaneously transmitted from multiple antennas, and the extension to the MIMO structure can calculate the distance of the target and the Doppler can also be in the azimuth. The direction of arrival (DOA) is obtained in the angular and elevation directions. A three-dimensional imaging radar is realized by combining an Orthogonal Frequency Division Multiplexing (OFDM) signal model with a Direction of Arrival (DOA) algorithm and a virtual antenna geometry (Fig. 2.34).

## 2.6 Application of Millimeter-Wave Radar

### *2.6.1 Application of Unmanned Aerial Vehicle (UAV)*

The millimeter-wave radar is a radar sensor that senses the object by transmitting and receiving microwaves. By using the frequency difference between the transmitted and received signals, the existence, velocity, direction, distance and angle of the object motion can be calculated by the formula. It has the ability to penetrate fog, smoke, dust and has a small size, high degree of integration and sensitive sensing ability. Hence, it can be applied for all weather types and times, and satisfy the requirements of the unmanned aerial platforms, helicopters and small airships. The application of millimeter-wave radar on drones mainly includes the millimeter wave radar altimeter and obstacle avoidance millimeter wave radar (Fig. 2.35).

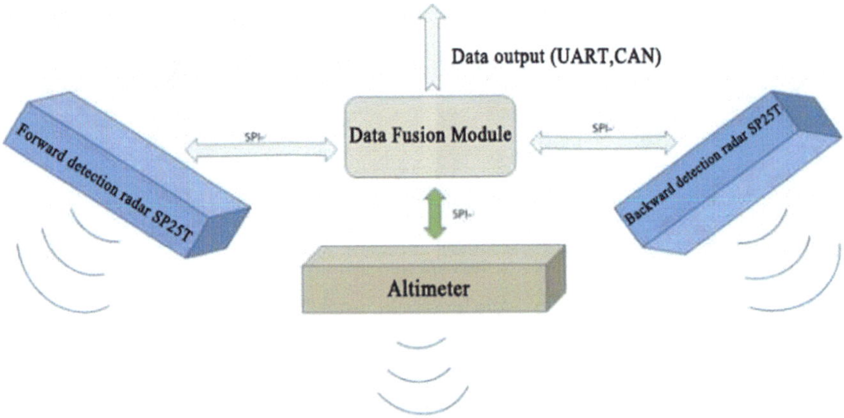


**Fig. 2.35** Framework of "1 + 2" radar system for unmanned drone

## 1. Millimeter-Wave Radar Altimeter

The drone height gauge is divided into two types: absolute height measurement altimeter and relative height measurement altimeter. The absolute height measuring instrument includes the GPS and the barometer, where the absolute height relative to sea level is obtained. The relative measurement is to measure the relative height of the UAV platform from the ground plane. It is generally used in low-altitude drone platforms. It can automatically set the height according to the terrain, adjust the height, and realize the imitation flight. The relative height measurement altimeter mainly includes high ultrasonic altimeter, laser and millimeter-wave radar altimeter. The millimeter-wave radar altimeter transmits millimeter wave for distance ranging which could penetrate fog, smoke and dust. It has a small body, high integration degree, sensitive sensing ability, and is independent of lighting which could achieve an all-weather and all-day application (Fig. 2.36).

A typical application of the millimeter-wave radar altimeter is on the plant protection UAV. Through the installation of the altimeter, the plant protection UAV could achieve about 1–2 m of fixed flight height above the crops, and is not affected

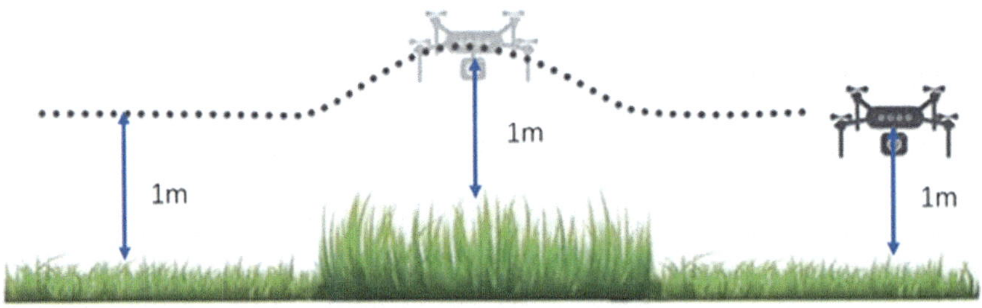


**Fig. 2.36** Schematic diagram of the millimeter-wave radar altimeter

by mist. At the same time, it achieves higher precision (the height error is less than ±10CM) which satisfies the distance requirements for pesticide spraying through uniform and efficient spraying.

2. **Millimeter-Wave Obstacle Avoidance Radar**

At present, there are automatic obstacle avoidance systems based on video images, ultrasonic radar or laser radar. However, these sensors have some defects. For example, the cost of the laser radar sensor is too high, and the ultrasonic sensor has a limited working distance and is easily interfered. The visual image sensor's detection performance under poor lighting is greatly reduced, and the requirements for the performance of the image processor is high. The millimeter-wave radar has a detection distance 5 times that of the ultrasonic wave, and the detection distance can reach up to 100 m. The small size, high resolution and low power consumption of the millimeter-wave radar fully meet the demanding requirements of the UAV for size and power consumption. At the same time, it has strong anti-interference ability and is little influenced by the weather. It can realize up to 32 targets which makes it suitable as an obstacle avoidance sensor.

After actual testing, the UAV equipped with the millimeter-wave obstacle avoidance radar can fly in the environment containing high voltage lines, buildings and roads. The millimeter-wave obstacle avoidance radar system can detect millimeter-level transmission cables well which takes care of the drone's safety issues (Fig. 2.37).

**Fig. 2.37** Application of the millimeter-wave obstacle avoidance radar

### 2.6.2 Unmanned Vehicle Application

In order to better describe the functions of the millimeter-wave automotive radar systems and specific applications, Fig. 2.38 could be referred to.

Millimeter-wave car radar sensors can be distributed around the vehicle body, such as the front and rear of the vehicle, the sides of the vehicle and the four corners of the vehicle. The millimeter-wave radar of different installation parts has different functions, which are mainly divided into three categories: adaptive cruise control (ACC), front and rear anti-collision system (F/RCW), and blind spot detection system (BSD) and parallel auxiliary system (LAC). Based on the different installation parts, it can further be divided into the parking assist system (PAS), cross traffic assist system (RCTA) and lateral anti-collision system (SCD), etc.

1. **Adaptive Cruise System**

The adaptive cruise system is generally installed in front of the vehicle and has a long working distance which belongs to the long-distance millimeter-wave radar with a detection range of more than 300 m. However, the radar viewing field is narrow and it mainly adjusts the distance and speed of the vehicle in relation to the object that is directly in front of the vehicle (generally refers to the vehicles in front within the same lane), thereby, ensuring the driving safety of the vehicle. When necessary, the system could make emergency braking to prevent collisions and achieve safe automatic cruising (Fig. 2.39).

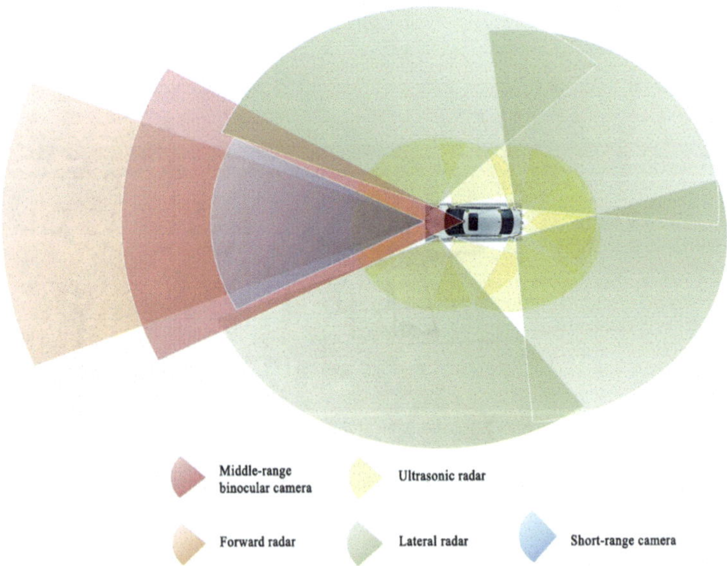


**Fig. 2.38** Self-driving car perception area

**Fig. 2.39** Adaptive cruise control system

## 2. **Anti-Collision System**

The anti-collision system generally has a working distance of within 100 m and belong to the medium range millimeter-wave radar. The system has extremely high detection accuracy requirements for positioning and monitoring purposes. In addition, the system also imposes strict requirements on object resolution and data update rate to ensure that the system can quickly and accurately obtain object information around the vehicle body so as to provide drivers with more accurate and efficient decision information to minimize the chance of a collision. Therefore, the anti-collision system is the crucial factor in ensuring the safe travel of unmanned vehicles in today's context. Anti-collision radars are generally installed on the front and rear of the vehicle body and on both sides of the vehicle body to ensure the safety of both the vehicle and the passengers in it (Fig. 2.40).

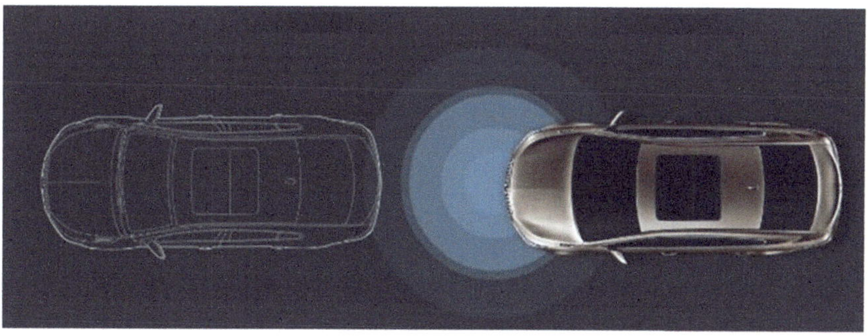

**Fig. 2.40** Anti-collision system

3. **Blind Spot Detection System and Parallel Auxiliary System**

The blind spot detection system and the parallel auxiliary system, that are generally installed in the blind spot area at the four corners of the vehicle body, belong to the short-range millimeter-wave vehicle radar. The working distance of the parallel auxiliary system is generally about 30 m while the blind spot detection system has an even shorter distance within 10 m. Other than the difference in the working distance, the installation positions and functions of the two systems are basically the same with the same purpose to provide the driver with information in the blind zone and to facilitate safe driving (Fig. 2.41).

## *2.6.3 Application on Unmanned Boats*

According to statistics, about 89–96% of the collisions involving boats, ships or vessels are attributed to human-induced causes which include both apparent and implicit causes. The anti-collision system based on the millimeter-wave radar used in the unmanned ships may be a good solution to address this problem. In the early years, collisions between ships, bridges or coasts occur frequently. To solve this issue, the use of tires on shore to prevent collision was implemented. Although many ships still use this method, the results are still unsatisfactory (Figs. 2.42 and 2.43).

The propagation speed of ultrasonic waves in water is 1400 m/s. The propagation speed in air is 340 m/s at 15 °C. The higher the temperature, the higher the propagation speed. Under normal circumstances, the propagation in solids is the

**Fig. 2.41** Blind spot detection system

(a) (b)

**Fig. 2.42** Ship collision; **a** collision between ships; **b** ship collision on shores

**Fig. 2.43** Tires landing anti-collision method

fastest, followed by liquids and then in gases. With respect to the situation of concurrently coming into contact with the water surface and implementing the reflection of the front object, the distance is often reported incorrectly. When the distance is too short, the ultrasonic solution is rarely used.

In the process of using the laser radar, the environment of the unmanned ship is often complicated. The laser can penetrate the water surface but the data obtained is not necessarily the actual target distance. In the marine unmanned system, no one has yet to use this solution.

Using the 1 + N number of millimeter-wave radar, that is to use one long-range millimeter-wave radar and N number of medium-range millimeter-wave radar, the long distance could be selected between 100–450 m to realize the forward anti-collision. According to the coverage area and the different sizes of the ships, N number of medium-range millimeter-wave radars could be selected based on the actual condition to achieve a 360° protection without dead angle (Fig. 2.44).

**Fig. 2.44** Millimeter-wave radar on unmanned ships

# Bibliography


1. Kraus JD (1988) Antennas, 2nd edn. McGraw-Hill, New York
2. Silver S (1949) In: Microwave antennas theory and design. MIT radiation laboratory series, vol 12. McGraw-Hill, New York
3. IEEE Standard Letter Designations for Radar-Frequency Bands, IEEE Std 521–1984
4. Shao Z, Zhu J et al (2001) Millimeter wave radar and millimeter wave phased array radar. The 14th of the ministry of information industry (2)
5. Jiang C, Zhang M, Jiao P (1997) Handbook of radar wave propagation and attenuation
6. Barton DK, Radar LSA (1997) Technology encyclopedia. Artech House Inc., Boston
7. Skolnik MI (1980) Introduction to radar systems, 2nd edn. McGraw-Hill, New York
8. Chinese Military Encyclopedia (1994) Electronic countermeasure and military radar technology volume. Military Science Press, Beijing
9. Zhou M (2007) Design and application testing system shipboard radar circuit. Electron Meas Technol 30(9):1,572,159
10. Liqin F, Zhiguo G et al (2004) Digital signal processing theory and implementation. National Defense Industry Press, Beijing
11. Xiang J (2001) Linear frequency modulation and binary coded radar signal. In: Xu Q, Xu J (eds) Radar system. Electronic Industry Press, Beijing
12. Morchin W (1993) Radar engineer's sourcebook. Artech House Inc., Boston
13. Bin W et al (2006) Research progress of vehicle ranging radar. Sens Microsyst 25(3):7–9
14. Zhang J (2001) Research on millimeter wave vehicle collision avoidance radar. Doctoral dissertation, Nanjing, Nanjing University of Science and Technology
15. Ding L, Geng F (1995) Radar principle. Xi'an, Xi'an University of Electronic Science and Technology Press
16. Jingxuan G, Qiuhui Z, Jinke Z, Youhong X, Wenqiang J, Lilong L (2015) Research on vehicle collision avoidance millimeter wave radar ranging system. Henan Sci Technol 04:1–3
17. Wang H, Zhao F, Deng Y (2015) Development and application of millimeter wave synthetic aperture radar. J Infrared Millimeter Waves 34(04):452–459
18. Yu L (2016) Millimeter wave near-field imaging technology based on MIMO array. Beijing Institute of Technology
19. Wang S, Dai X, Xu N, Zhang P (2017) Overview of the driverless car environment perception technology. Changchun University of Technology (Natural Science) 40(01):1–6
20. Zhao S (2010) Radar signal processing technology. Tsinghua University Press, Beijing, p 2
21. Fu Z (2016) Research on beamforming in millimeter wave MIMO system. University of Electronic Science and Technology

22. Li F (2015) Bayesian compressed sensing theory and technology. University of Electronic Science and Technology
23. Li F, Guo Y (2015) Analysis of compressed sensing. Science Press, Beijing, pp 62–79
24. Heuel S, Rohling H (2011) Pedestrian classification in automotive radar systems. Radar symposium. IEEE, pp 477–484
25. Heuel S, Rohling H (2011) Two-stage pedestrian classification in automotive radar systems. Radar symposium. IEEE, pp 1–8
26. Heuer M, Al-Hamadi A, Rain A, et al (2014) Pedestrian tracking with occlusion using a 24 GHz automotive radar. Radar symposium. IEEE, pp 1–4
27. Wu S, Mei X et al (2008) Radar signal processing and data processing technology. Publishing House of Electronics Industry, Beijing
28. Hu K, Hu A (2006) Application of digital beamforming technology (DBF) in radar. Mod Defense Technol 34(6):103–106
29. Chen W, Bi X, Cao Y (2011) Millimeter-wave automotive collision avoidance radar waveform design. Comput Meas Control 19(11):2714–2716
30. Liu L, Wang W, Wei Z et al (2016) Millimeter wave research unmanned helicopter obstacle avoidance radar system. Naval Aeronaut Eng Inst 31(2):143–146
31. Bi X (2012) Research on several key issues of radar detection technology in traffic safety field. University of Chinese Academy of Sciences; Graduate University of Chinese Academy of Sciences
32. Wu Y (2009) Research on key technologies in early warning system of automobile collision avoidance radar. University of Electronic Science and Technology of China
33. Xiao H, Yang J, Xiong J (2005) LFMCW signal utilizing its radar multi-target MTD-speed matching method. J Radio Sci (6th edition)
34. Wang Y, Chen H et al (2004) Theory and algorithm of spatial spectrum estimation. Tsinghua University Press, Beijing, pp 406–407
35. Chen J, Gu H, Su W (2009) A fast multi-target positioning method for bistatic MIMO radar. J Electron Inf Technol 31(7):1664–1668
36. Skolnik MI (2008) Radar handbook (3rd edn). McGraw Hill, New York
37. Zwanetski A, Kronauge M, Rohling H (2013) Waveform design for FMCW MIMO radar based on frequency division. Radar symposium. IEEE, pp 89–94
38. Rohling H, Kronauge M (2015) New radar waveform based on a chirp sequence. Radar conference. IEEE, pp 1–4
39. Zwanetski A, Rohling H (2012) Continuous wave MIMO radar based on time division multiplexing. Radar symposium. IEEE, pp 119–121

# Chapter 3
# LiDAR Technology

## 3.1 Introduction

LiDAR, also known as light detection and ranging, has seen rapid development ever since people started to first use laser for optical distance measurement. In the field of military application, for instance ranging and weapon guidance, multi-spectral laser imaging systems with high range resolution, single photon sensitive arrays, and wide emission spectral sections have been intensively developed. These systems are immune to weather and can penetrate through foliage, clothes and dense media for target recognition, and optical coherence tomography of the building to reconstruct its three-dimensional structure. Multi-dimensional measurements are primarily limited by laser power, computational processing efficiency, incoherent and coherent focal plane arrays, and signal processors. In the field of civilian application, advanced LiDAR is used to measure the optical parameters and elements of the wind, temperature, atmosphere and cloud, as well as, sea ice and sea surface measurements for the environmental research. Imaging and mapping are also important areas of LiDAR applications. Amongst the different types of LiDAR, the synthetic aperture LiDAR has a long development cycle and has been tested in principle-based experiments. For instance, coherent LiDAR, azimuth and angle measurement LiDAR, multi-spectral LiDAR, etc., play an important role in the practical application of precise speed, angle and spectrum measurement. Micro-laser is also widely used in ophthalmology research and treatment because it can map the refractive properties of the human eye.

LiDAR can be classified based on the different methods. According to the transmission waveform and data processing methods, it can be classified into the pulsed LiDAR, continuous wave LiDAR, pulse compression LiDAR, moving target display LiDAR, pulse Doppler LiDAR and imaging LiDAR. Based on installation platform, it can be further classified into ground LiDAR, airborne LiDAR, shipborne LiDAR and aerospace LiDAR. According to the mission purpose, it can be further classified into the fire control LiDAR, ranging LiDAR, missile guidance

 
X. Bi, *Environmental Perception Technology for Unmanned Systems*, Unmanned System Technologies, https://doi.org/10.1007/978-981-15-8093-2_3

LiDAR, obstacle avoidance LiDAR and aircraft landing guidance LiDAR, etc. In the past decade, space-based LiDAR technology for remote sensing has been rapidly developed on a global scale with the application of research increasing in depth and breadth. In the various aspects such as the atmospheric environment, topographic mapping, urban management, disaster monitoring and heritage protection it has played an important role. LiDAR technology provides the three-dimensional topographic maps and high-precision measurement of the distance and approaching speed between the spacecraft and the planet's surface. This allows the large robots and spacecrafts to land on the planet accurately and safely. Hence, National Aeronautics and Space Administration (NASA) and European Space Agency (ESA) had adopted LiDAR technology as one of the key enabling technologies to achieve the automated and safe landing of the future space robots and spacecrafts.

## 3.2 Concept and Characteristics of LiDAR

### *3.2.1 Laser*

Light Amplification by Stimulated Emission of Radiation (Laser) refers to the expansion of the stimulated emission of light, as the name indicates, summarizes the fundamental principles of laser. The following section briefly describes the conditions to generate laser.

We are aware that the microscopic particles have a specific energy level (usually discrete) and that these microscopic particles can interact with light which is shown in Fig. 3.1. When a particle transits from one energy level to another, it can either absorb or radiate the photons accordingly. The energy of the photon is the energy difference between the two energy levels.

As laser does not originate from nature but is a product of the human-assisted process of light stimulation, it therefore encompasses the following characteristics that are missing from normal light:

1. Strong directionality. The divergence angle of the laser beam is very small, generally about 0.18°. This is in 2–3 orders of magnitude smaller than ordinary light and microwave. Therefore, the solid angle is extremely small and can generally be as small as $10^{-8}$ rad.
2. Good monochromatic property. The narrower the spectrum from the emitting light source, the better is the monochromatic property.
3. High brightness. As the laser energy is highly concentrated in space, its brightness is millions of times higher than that of ordinary light sources.
4. Good coherence. The coherence of light means that when two beams meet, the waves in the encounter area are superimposed, and a clear interference pattern can be formed or a stable beat signal can be received. The light emitted by the same light source at different times within the coherence time, t, will interfere

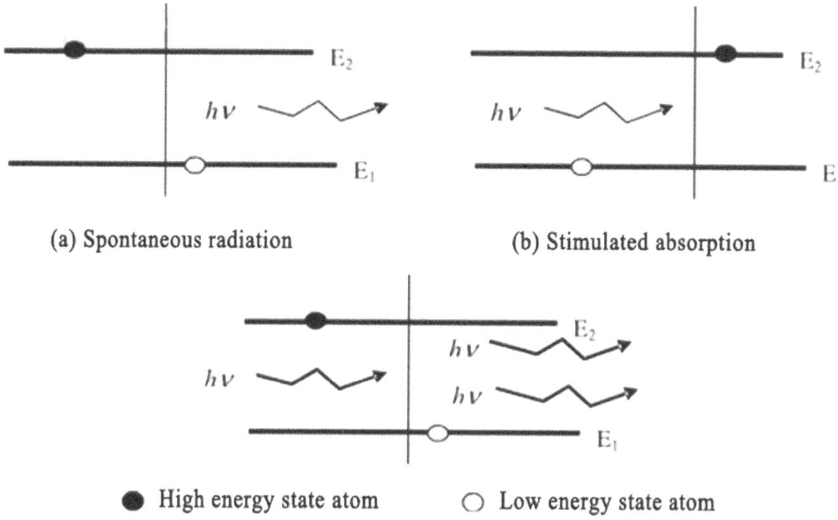


**Fig. 3.1** Distribution of one atom energy level

when it meets through different paths. This coherence is called temporal coherence. At the same time, the coherence of light emitted by different points in space is called spatial coherence. The laser is formed by stimulated radiation. The propagation direction, vibration direction, frequency and phase of the light waves emitted by each illuminating center are completely consistent, so the laser has good temporal and spatial coherence.

### 3.2.2 LiDAR Function and Principle

Light Detection and Ranging (LiDAR) is an optical measuring method for locating and measuring the distance to an object in space. In principle, the system is similar to the radar method, but the laser does not use microwaves. Instead, it uses ultraviolet, infrared or visible light.

The duration when the light (laser) pulse is emitted until the reflected light is received is proportional to the distance between the detection system and the detected object. When the speed of light is approximately 3,000,000 km/s (in the air) and the distance is 50 m, the duration of measurement is about $3 \times 10^{-7}$ s or 333 ns. This could be seen in Eq. (3.1).

$$r = \frac{c \cdot t_{of}}{2} \tag{3.1}$$

where $r$ is distance (m); $c$ is the speed of light ($3 \times 10^{8}\,\mathrm{m/s}$); $t_{of}$ is the duration (s).

Since the emitted light pulse must pass the distance between the transmitter and the receiver twice, the time, $t_{of}$, represents twice the distance between the object.

If there are multiple objects in one measurement channel, multiple objects can also be acquired when using the corresponding evaluation method. If the attenuation of the atmosphere increases (for example, due to fog), the single pulse will reflect from the water droplets in the air. Depending on the optical design of the radar system, this can cause the receiver to saturate such that measurements can no longer be taken.

However, today's sensors have the ability to dynamically match sensitivity, and together with the multi-target capability within the measurement channel, the sensors can be used to measure soft atmospheric disturbances in response to climatic objects. At the same time, it is also helpful to receive different distances based on the different thicknesses of the fog. Combining the individual impulse responses of a longer measurement pitch (time window) results in a flat, enlarged echo.

Range performance is affected by the intensity of the emitted light pulse and the sensitivity of the receiver. At the same time, pulse power is limited by visual safety requirements. In contrast, other parameters such as atmospheric transmission, object size or reflectivity are not affected.

In the case where the beam radiating surface is smaller than the object, the received signal strength is given as:

$$P_r = \frac{KK \cdot A_t \cdot H \cdot T^2 \cdot P_t}{\pi^2 \cdot R^3 \cdot (Q_v/4) \cdot (\Phi/2)^2} \tag{3.2}$$

In the case where the target (the distance is large) is smaller than the beam radiating surface, the following relationship applies:

$$P_r = \frac{KK \cdot A_t \cdot H \cdot T^2 \cdot P_t}{\pi^2 \cdot R^4 \cdot (Q_v \cdot Q_h/4) \cdot (\Phi/2)^2} \tag{3.3}$$

Wherein, $P_r$ is the received signal strength (W); $KK$ is the reflection of the measured object; $\Phi$ is the reflected angle of the object (radians); $H$ is the object width (m); $T^2$ is the atmospheric transmittance; $Q_v$ is the elevation angle (radians); $Q_h$ is the azimuth angle (radians); $A_t$ is the receiving lens area ($m^2$); $P_t$ is the laser power (W).

### 3.2.3 LiDAR Characteristics

LiDAR is a combination of the laser and the radar technology. By using the laser as the emitter of the radar, the laser's high collimation, monochromaticity and high strength can be fully utilized. Moreover, the wavelength of the laser is short, which is more than three orders of magnitude shorter than the wavelength of the microwave, and thus the laser radar is widely used. In comparison with the microwave radar, laser radar has the following advantages:

1. High angular resolution. According to Rayleigh criterion:

$$\sin\theta = 1.22\frac{\lambda}{D} \tag{3.4}$$

where D is the aperture of the optical receiving system. It can be observed that as the wavelength of the laser is short, when the aperture of the optical receiving system is small, a higher angular resolution is obtained.

2. The high-speed resolution means that the LiDAR has a wider range of speed measurement. When the detected object has a relative speed to the LiDAR, the Doppler shift $f_d$ expression is given as:

$$f_d = \frac{2V_r}{\lambda} \tag{3.5}$$

Wherein, $V_r$ is the radial velocity between the measured object and the laser radar. Due to the short wavelength of the laser, the Doppler frequency sensitivity is high, and hence, the speed resolution of the laser radar is high.

3. High distance resolution

The laser speed measurement formula is as follows:

$$r = \frac{c \cdot t_{of}}{2n} \tag{3.6}$$

Wherein, R is the measured distance, $\Delta t$ is the flight time of the laser pulse, c is the speed of light in vacuum, and n is the refractive index of the transmission medium. Since the pulse width of the laser can reach the order of picoseconds, the energy of the laser is concentrated, so the distance resolution is increased to the order of millimeters.

In comparison with the millimeter-wave radar, in terms of accuracy, the detection range of the laser radar will not be restricted by band loss (the millimeter-wave radar must use high-bandwidth radar for far detection) which could detect objects such as the pedestrians and could perform an accurate modelling of the surrounding obstacles. However, the disadvantage of the LiDAR is that it is affected by the weather and atmospheric conditions. Under clear weather condition, the laser generally has a small attenuation and long propagation distance. Under inclement weather conditions such as heavy rain, dense haze and fog, the attenuation will increase sharply and the propagation distance will be greatly affected. In contrast, the millimeter-wave radars have a strong ability to penetrate fog, smoke, and dust,

**Table 3.1** Comparison between LiDAR and millimeter-wave radar

| | LiDAR | Millimeter-wave radar |
|---|---|---|
| Detection range | +++ | +++ |
| Field of vision | +++ | ++ |
| Width and height measurement | +++ | – |
| 3D modelling | +++ | – |
| Long range target detection | +++ | – |
| Accuracy | +++ | – |
| Rain, snow, dust | ++ | +++ |
| Fog | + | +++ |
| Night detection | +++ | +++ |
| Marked object and colour recognition | + | – |
| Cost | High | Low |

so detection could still be done under extreme weather conditions. Table 3.1 compares the characteristics of LiDAR and millimeter wave radar.

## 3.3 LiDAR Analysis

### 3.3.1 Mechanical LiDAR

The mechanical scanning method is performed by rotating at 360° or other larger angles via motor-driven single-point or multi-point ranging modules. This method is the most direct and the technical difficulty is the lowest. Hence, mechanical scanning LiDAR is the first to be applied (Fig. 3.2).

However, considering the factors such as lenses, mechanical structures, and circuit boards, multi-point ranging modules are usually not optimized in terms of size and weight. Therefore, when the motor rotates the module for a long period of time, the bearings are easily lost which result in increased costs due to reduced reliability and increased losses.

### 3.3.2 Solid State Hybrid LiDAR

The mechanical LiDAR will rotate 360° at all times during operation. When the hybrid solid-state LiDAR is in operation, the rotation could not be seen from the outside. The sophistication is that the mechanical rotating parts are made smaller and deeply hidden in the outer casing.

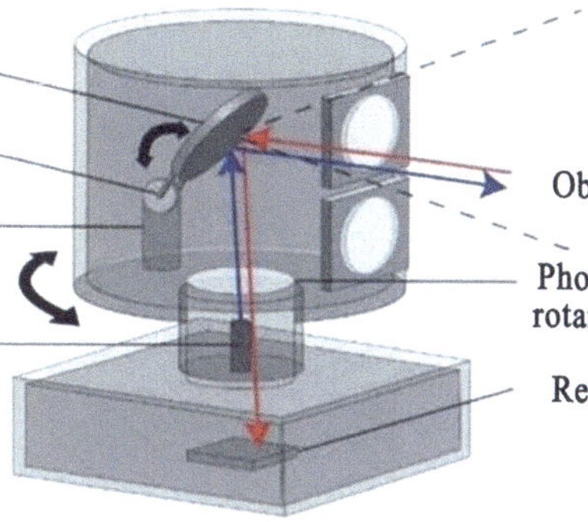


**Fig. 3.2** Illustration of a typical LiDAR optical scanner

The Solid-State Hybrid LiDAR conducts point measurement of reflectivity and distance based on a 360° horizontal field of vision and up to 40° of vertical field of vision, updating at 20 times every second. The Solid-State Hybrid LiDAR uses the transit time (TOF) for measurement, that is, light pulses are emitted to measure the object's movement. In addition, the speed of light is used to accurately measure distance. In the case of the Solid-State Hybrid LiDAR, multiple light pulses from multiple lasers simultaneously pulse and measure the distance in nanoseconds. The formula that is used to measure the distance for a returned photon to reach and then leave an object is given as:

$$d = c \times t_{tof} \tag{3.7}$$

where d is the distance; c is the speed of light; $t_{tof}$ is the time of flight.

Conventional scanning imaging LiDAR systems generally use a double-swing mirror, a double galvanometer, and a rotating polyhedral reflective prism. The scanning system that consists these macro-sized optical components is bulky and cumbersome. By using the Micro-Electro-Mechanical System (MEMS), the silicon chip could directly be integrated to form a delicate micro-scanning mirror. Through the micro-scanning mirror, the laser light is reflected and thereby achieve a micron scanning motion, and thus, any mechanical rotating parts in the LiDAR are not visible macroscopically (Fig. 3.3).

Hybrid solid-state LiDAR uses a MEMS scanning mirror which only needs a laser source to reflect the laser light through a MEMS scanning mirror. The two work together in a microsecond frequency to achieve the target 3D scanning after receiving by the detector. Compared to the mechanical LiDAR structure of multiple sets of chipsets, the hybrid solid-state LiDAR uses a single set of MEMS scanning mirrors and a single beam laser source, which yields obvious volume reduction (Fig. 3.4).

Figure 3.5 illustrates a typical MEMS LiDAR system structure which uses the MEMS micro mirror to perform scanning at the transmitter and linear array

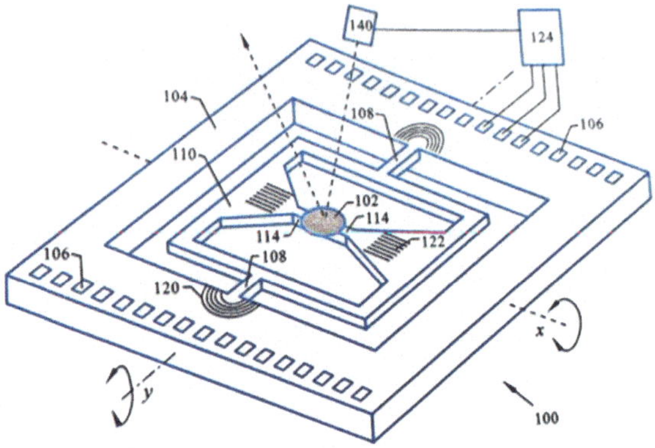


**Fig. 3.3** Schematic diagram of the MEMS scanning mirror

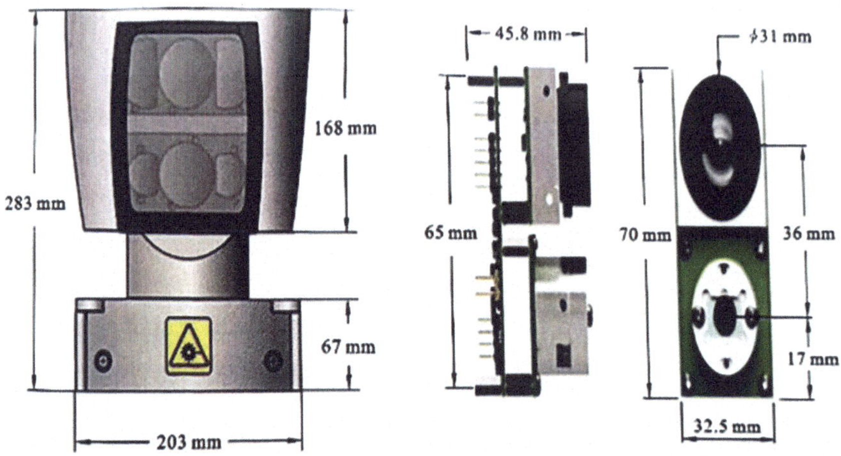


**Fig. 3.4** Comparison of mechanical LiDAR and hybrid solid-state LiDAR dimensions

receivers at the receiving end. Hence the MEMS program caters for only one option of scanning at the transmitter end, and the issue at the receiving end is left to the system developer to resolve. The main advantage of using MEMS LiDAR is in its speed of landing. The micro mirror technology is relatively mature and hence procurement is easily achieved with different suppliers. As the MEMS micro mirror is generally very small (from smaller than a millimeter to several millimeters), a large shock to the micro mirror would reduce the frequency significantly, and hence it is not suitable for fast scanning purpose. Another limitation of the MEMS LiDAR is that under the condition of complicated optical paths, for example beam expansion, the lenses must be expanded for large angle scanning since the micro mirror scanning angle is relatively small (from a few degrees to 60°).

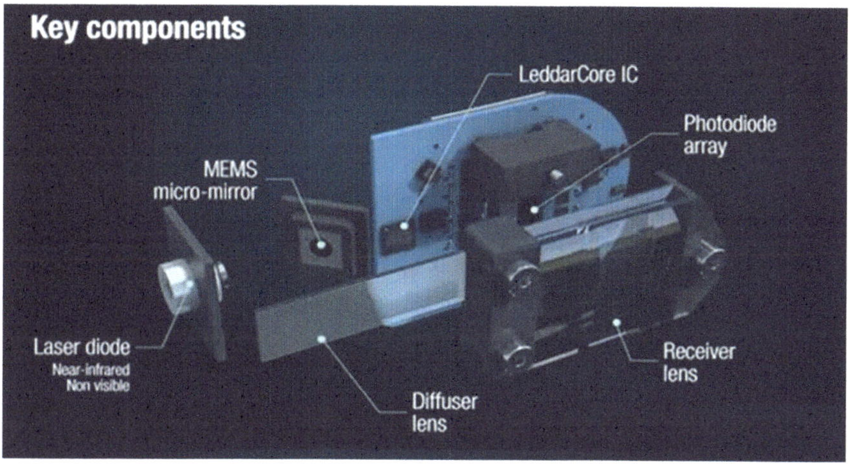


**Fig. 3.5** Hybrid solid-state LiDAR structure

Photodetectors are components that convert light pulses into electrical signals and act as "eyes" in LiDAR systems. At present, the main photodetectors are Avalanche Photon Diode (APD), Single Photon Avalanche Diode (SPAD), Silicon Photomultiplier (MPPC) and PIN Photodiode.

1. **APD/SPAD**

The working mode of APD is divided into linear mode and Geiger mode. When the bias voltage of the APD is lower than its avalanche voltage, it linearly amplifies the incident photoelectrons. This operating state is called the linear mode. In linear mode, the higher the reverse voltage, the greater the gain. The APD performs equal gain amplification on the input photoelectrons to form a continuous current, and obtains a laser of continuous echo signal with time information. When the bias voltage is higher than its avalanche voltage, the APD gain increases rapidly, and the single photon absorption can saturate the detector output current. This working state is called Geiger mode. The APD that works in Geiger mode is also called SPAD (Fig. 3.6).

For an APD operating under the Geiger mode, a single photon can make its working state switch between on and off to form a steep echo pulse signal, and thus explains its ability of single photon imaging. The sensitivity of this kind of photodetector is extremely high and hence its detection distance theoretically could reach extremely far distance, that is, 3000 km would not be a problem. Therefore, it had already made great achievements in the military field (stealth aircraft and missile system) a few years ago. Hence, the AGP's Geiger mode is ideal for use in LiDAR.

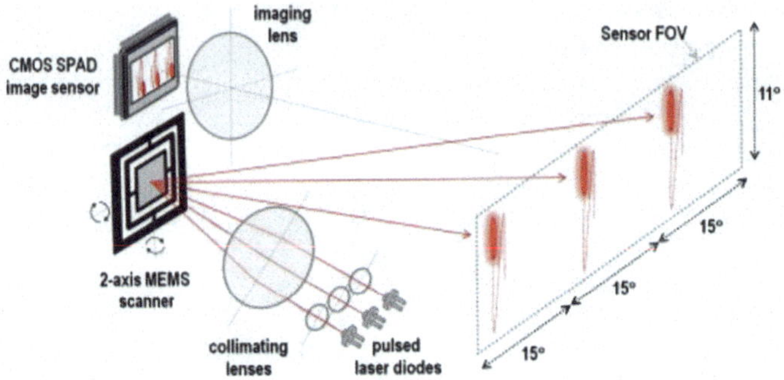


**Fig. 3.6** Application of 2-axis MEMS scanning mirror and SPAD hybrid solid-state LiDAR

2. **MPPC**

MPPC is a new type of optical semiconductor device that is commonly known as the Silicon Photomultiplier (SiPM). According to its principle, it can also be called the Multi-Pixel Photon Counter (MPPC). It consists of multiple APD arrays operating in Geiger mode with high gain, high detection efficiency, fast response, excellent time resolution and wide spectral response range.

When a pixel in the MPPC receives an incident photon, it outputs a pulse of a certain amplitude. If multiple pixels receive the incident photons, each pixel will output a pulse, which will eventually be superimposed and output by a common output to achieve greater gain.

Compared with the APD, the gain of MPPC can reach 10^5 to 10^6, so in theory, longer distance information can be obtained in a shorter time, and the detection bandwidth is also comparable to APD. Additionally, with the small effective area of the pixel structure, the MPPC is not only able to provide faster time characteristics (rise time of only about 1 ns), but also with its unique ability to distinguish the different reflectivity of object surfaces, it can achieve the purpose of both ranging and distinguishing surface characteristics of an object at the same time.

### *3.3.3 Solid State LiDAR*

In recent years, new LiDARs based on the unmanned systems have begun to show trends of solidification, miniaturization and low cost. Solid-state technology is undoubtedly an important technical direction for the development of new LiDARs. Pure solid-state laser radar is generally classified into optical phased array LiDAR and area array imaging LiDAR.

1. **Optical Phased Array (OPA)**

OPA LiDAR adopts a phased array design. It is equipped with an array of transmitters which adjust the relative phase of the signal to change the emission direction of the laser beam, so as to achieve the purpose of optical scanning.

Phased array radar is mainly composed of a laser transmitter, a receiver, a programmable optical phased array beam controller, an optical system, integrated information processing unit and the integrated display units. The optical transmitter and receiver channels are separated from each other but share an optical aperture to ensure that parallax is not generated at close range (Fig. 3.7).

Phased array LiDAR is an active imaging system that uses a near-infrared wavelength laser as a detection carrier, and emits a modulated laser beam to illuminate the target to be measured. The target detects the reflected signal of the laser to determine the distance and reflection intensity of the target. The laser radar adopts the direct detection method, and uses the multi-sensor and programmable optical phased array scanning technology to realize the imaging of the region, and obtains the detailed information of the target through the interpretation of the image.

Optical phased array LiDAR can be further subdivided into electro-optic scanning and acousto-optic scanning according to the specific method of optical phase control. At present, the electro-optical scanning is mainly adopted which largely focused on the design and improvement of the phased array scanning components. In general, the main areas include the following: (1) array of optical waveguides; (2) fiber grating; (3) liquid crystal material; (4) electro-optical scanner based on ferroelectric domain (Table 3.2).

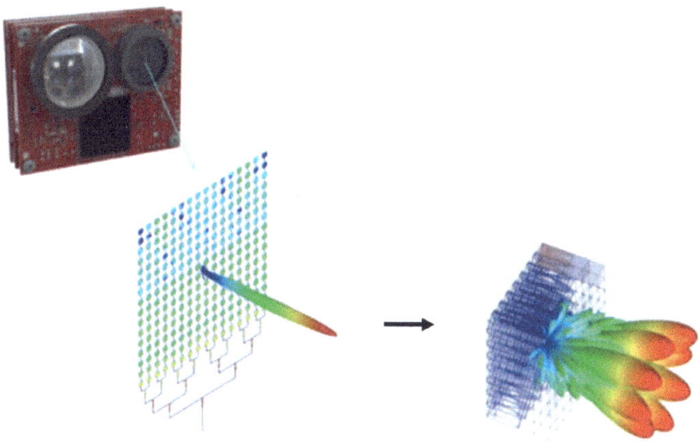

**Fig. 3.7** Schematic of OPA LiDar scanning concept

**Table 3.2** Comparison of the different optical characteristics of phased array mode

| | Optical phased array | | | |
|---|---|---|---|---|
| | Lithium niobate crystal | PLZT piezoelectric ceramics | Liquid crystal | Optical waveguide array |
| Response time | ps | ns | ms | ns |
| Driving voltage | <10 | 1000 | <10 | <10 |
| Scanning angle (°) | <0.1 | <0.1 | <10 | ≈30 |

2. **No-Scan 3D Imaging LiDAR (Area Array Imaging LiDAR)**

No-scan (also known as non-scanning) three-dimensional imaging LiDAR has a large field of vision, fast imaging speed, extensive distance detection, and high-resolution advantages. At present, there are two different types of implementation of the non-scanning three-dimensional imaging LiDAR technology. The first one is to directly acquire the three-dimensional image of the target by using the three-dimensional imaging sensor, while the other is to synthesize the intensity image and the distance image by using the two-dimensional imaging sensor. The specific implementation mainly includes the following methods.

(1) Three-dimensional imaging LiDAR based on area-free focal plane detector without scanning the direct time-of-flight ranging method

The detector integrates a high-precision time difference measurement circuit behind each detector cell. Each pixel of the detector acts as a separate detector that outputs a separate signal. Subsequently, the circuit measures the time difference between the time the laser is emitted and the time when the reflected laser is received, and obtained the corresponding distance of each pixel. Hence, just by emitting one laser pulse, the entire three-dimensional image of the object illuminated by the laser can be obtained. Furthermore, the imaging speed is fast. FLASH laser imaging radar is a representative product which adopts a key technology of the FLASH sensor development. Currently, larger area array of time-measuring avalanche diode array (APD) area array detectors and the Geiger mode avalanche diode array (GM-APD) area array detectors have been developed abroad (Fig. 3.8).

(2) Three-dimensional imaging LiDAR based on scan-free indirect time-of-flight ranging method based on dedicated modulation and demodulation area array detector

The difference between the device and the ordinary image sensor is that, firstly, it has a high-speed shutter and can achieve an exposure speed of 10 ns. Secondly, it has an exposure accumulation function, which can accumulate signals generated by successive multiple exposures. Distance ranging method adopts the indirect time-of-flight distance method which is based on the cosine wave-square wave

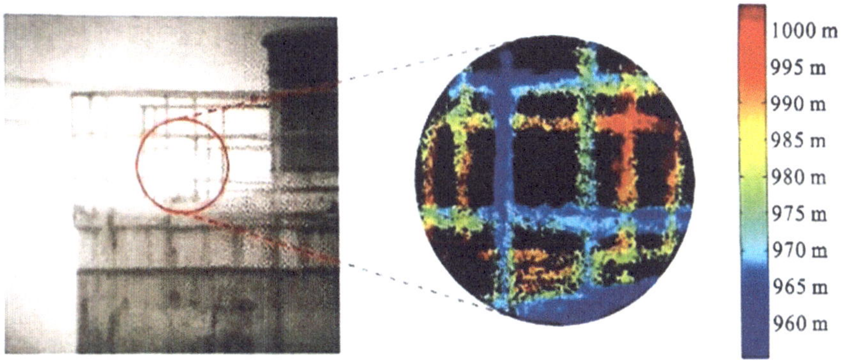


**Fig. 3.8** Structure and performance visualisation of area array LiDAR

phase method. In the subsequent circuit calculation of the photon mixed-frequency detector, the phase, amplitude and offset of each pixel corresponding to the incident light could be directly calculated. Henceforth, the distance and grey value of each point are calculated to obtain the depth map and the grey-scale image. At present, relatively mature modulation and demodulation area array detectors have been developed abroad, but the detector has a short integration time and low detection sensitivity. As a result, the imaging distance is only of a dozen meters, and could not realize long-distance three-dimensional imaging.

(3) Three-dimensional imaging LiDAR based on ICCD scan-free indirect time-of-flight ranging method

The initial non-scanning three-dimensional imaging technologies are based on the ICCD sensor device which performs waveform modulation on cosine wave, square wave or triangular wave for the light source and image intensifier gain according to the indirect time-of-flight ranging theory to obtain the three-dimensional information. Representative research institutions for this technology are the US Sandia National Laboratory and the Japan Broadcasting Corporation. There are also a small number of research institutions in China that have conducted research in this area. Due to the low frame rate of ICCD and low energy utilization of the light source, the imaging technology also has the disadvantages of low imaging speed and close proximity.

As illustrated in Fig. 3.9, the current market or on-going research on the area array LiDAR technology mainly include the following categories. The imaging characteristics of each technical solution are different for different application scenarios.

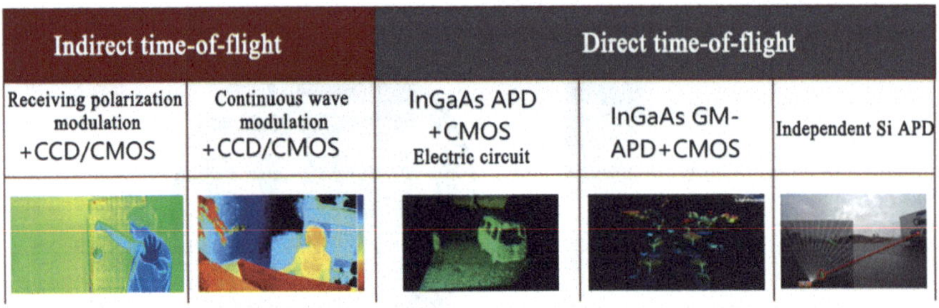


**Fig. 3.9** Area array technology solutions and output images

# 3.4 LiDAR Technology

## *3.4.1 LiDAR Signal Processing Technology*

Linear modulated signal, also known as chirp signal, refers to the linear increase or decrease in the frequency of a signal over time, or the linear decrease first before the increase of the signal frequency as shown in the diagram below. It has been widely used in many fields such as radar, wireless communication, medical imaging and modern instrumentation. Thus, it is closely monitored and extensively studied by local and international researchers.

The mathematical expression of the chirp signal can be described by the following formula:

$$s(t) = rect(\frac{t}{T})e^{j2\pi(\frac{k}{2}t^2 + f_c t)} \quad 0 \le t \le T \tag{3.8}$$

In the formula, $k$ is the frequency modulation slope (chirp rate), $f_c$ is the carrier frequency, $T$ is the pulse time width, and $rect(t/T)$ is the rectangular signal.

$$rect(\frac{t}{T}) = \begin{cases} 1, & |\frac{t}{T}| \le 1 \\ 0, & elsewise \end{cases} \tag{3.9}$$

Its instantaneous frequency is the first derivative of the phase term in the above equation and hence is given as:

$$f_i(t) = \frac{1}{2\pi}\frac{\mathrm{d}\theta(t)}{\mathrm{d}t} = kt + f_c \tag{3.10}$$

The bandwidth of the signal is $B = kT$. The time-bandwidth product of the chirp signal can be expressed as $BT = kT^2$.

When $k > 0$, the chirp rate is positive as shown in Fig. 3.10a. As such, the signal is up-chirped where the frequency increases linearly with time variation. The waveform of the signal has a rectangular envelope and varies linearly from sparse to

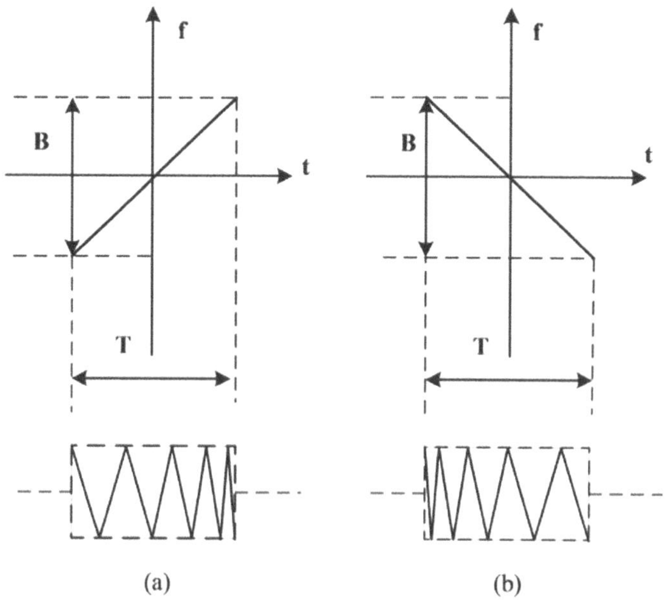


**Fig. 3.10** Illustration of linear modulated signal frequency and waveform. **a** Positive chirp rate; **b** negative chirp rate

dense. When $k < 0$, the chirp rate is negative as shown in Fig. 3.10b. The signal is down-chirped and its frequency decreases linearly with time. The waveform of the signal has a rectangular envelope and varies linearly from dense to sparse.

Traditional chirp signals are generated using electronics. Their performance is restricted by a limited sampling rate, and the bandwidth at which waveforms are generated is also limited. The use of optical technology to produce a chirp signal can overcome the above limitations. Compared with electronic technology, it has the advantages of wide waveform bandwidth, low phase noise, and easy tuning.

It is well known that when a very narrow time-domain light pulse is processed by an optical filter, the spectral characteristics of the light pulse usually change, that is, the spectrum of the light pulse is shaped by the optical filter. Obviously, if the time-domain width of the light pulse is very narrow (can be regarded as an impulse signal), the spectrum will be very wide. As a result, the optical signal after being processed by the filter will be closer to the impulse response of the filter. Additionally, the spectrum of the signal is also closer to the frequency response of the optical filter. Therefore, if a filter with a specific spectral amplitude response is used to spectrally shape a narrow optical pulse, the spectral shape of the output optical signal is approximately determined by the characteristics of the filter amplitude response. If the optical pulse that had been shaped by the filter goes through a dispersive medium with an appropriate amount of dispersion, the so-called frequency time-to-mapping effect will occur. The result of this effect is due to the mapping of the frequency domain amplitude characteristics of the shaped optical signal onto the amplitude characteristics of its time domain. That is, the time domain waveform of the output optical signal after frequency–time mapping is proportional to its frequency domain amplitude characteristic. Finally, the photo-electrical conversion is performed by a photodetector, and theoretically, a radio frequency signal proportional to the filter impulse response is obtained.

As shown in Fig. 3.11a, the direction of the arrow indicates the transmission direction of the optical and electrical signals, of which, the solid line represents the optical signal and the dotted line represents the electrical signal. It can be seen that the system is mainly composed of a pulse light source, an optical spectrum filter, a dispersive element and a photodetector. Of which, the pulse light source generates a narrow pulse signal in the time domain, and the active mode- or the passive mode-locked laser is commonly used; the function of the optical spectrum filter is to shape the spectrum of the light pulse; the dispersive element is used to generate the frequency–time mapping effect. It is usually implemented as a single mode fiber or fiber grating. The function of the photodiode is to achieve a photoelectric conversion of the signal to produce a signal of a target waveform.

Assuming the time width as $\Delta t_0$, the pulse of finite transformation as $x(t)$ and the spectral shaper shock response as $h(t)$, the optical pulse that is output by the filter can be expressed as:

$$y(t) = x(t) * h(t) \tag{3.11}$$

The corresponding Fourier transformation is given as:

$$Y(f) = X(f) \cdot H(f) \tag{3.12}$$

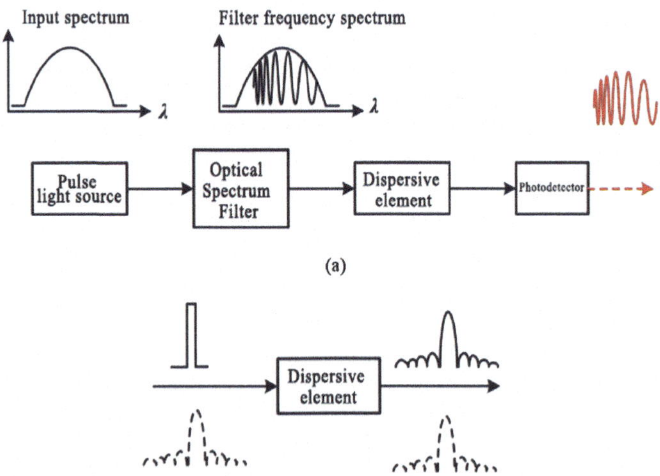


**Fig. 3.11** Based on the principles of frequency-time mapping of spectrum shaping to generate a chirp signal. **a** Principle map of spectral shaping and binding system; **b** principle of frequency-time mapping due to dispersion

Wherein, $X(f)$ and $H(f)$ are respectively the Fourier transform of the input light pulse $x(t)$ and $h(t)$. Since the input light pulse is narrow, the spectrum of the output optical signal $Y(f)$ has the characteristics of the spectral shaper frequency response.

Subsequently, the outputted optical signal propagates through the dispersive element. In general, common dispersive device can be regarded as a linear time-invariant system, and its frequency transfer function can be expressed as:

$$H(\omega) = |H(\omega)|\exp(-j\phi(\omega)) \tag{3.13}$$

Wherein, $H(\omega)$ is the amplitude response of the dispersive element which is regarded as a constant in theoretical analyses. The frequency–time function of the dispersive element mainly depends on the phase response of the dispersive element. Applying the Taylor series expansion, we can obtain:

$$\phi(\omega)=\phi_0+\dot{\phi}_0(\omega-\omega_0)+\frac{1}{2}\ddot{\phi}_0(\omega-\omega_0)^2+\frac{1}{6}\dddot{\phi}_0(\omega-\omega_0)^3 \tag{3.14}$$

where $\phi_0$, $\dot{\phi}_0$, $\ddot{\phi}_0$ and $\dddot{\phi}_0$ are respectively expressed as the phase constant of the carrier frequency, the group delay, the second-order dispersion coefficient (GVD) and the third-order dispersion coefficient (TOD). In order to simplify the analysis, in the frequency–time mapping analysis 为了简化分析, generally $\phi_0$, $\dot{\phi}_0$ 和 $\dddot{\phi}_0$ are ignored. Hence the above equation could be written as:

$$H(\omega) = |H(\omega)|\exp(-j\frac{1}{2}\ddot{\phi}_0\omega^2) \tag{3.15}$$

At this stage, the optical signal after transmitting through the dispersive element (satisfying the time domain condition $|\Delta t_0^2/\phi| << 1$) could be expressed as:

$$\begin{aligned} z(t) &= y(t) * \exp(j\frac{t^2}{2\ddot{\phi}_0}) = \int_{-\infty}^{\infty} y(\tau)\cdot\exp[j\frac{(t-\tau)^2}{2\ddot{\phi}_0}]d\tau \\ &\approx \exp(j\frac{t^2}{2\ddot{\phi}_0})\cdot\int_{-\infty}^{\infty} y(\tau)\cdot\exp[-j(\frac{t}{\ddot{\phi}_0})]d\tau = \exp(j\frac{t^2}{2\ddot{\phi}_0})\cdot Y(f)|_{f=\frac{t}{\phi}} \end{aligned} \tag{3.16}$$

The above equation shows that the intensity of the output signal is the Fourier transform of the envelope of the input optical pulse signal. This means that the spectral shape of the input light pulse can be mapped onto the pulsed time domain envelope stretched by the dispersion medium, and its corresponding mapping relationship can be expressed as $\omega = t/\phi$.

As seen from the above derivation, when the domain width of the light pulse and the dispersive element satisfy the condition $|\Delta t_0^2/\phi| << 1$, the outputted signal waveform amplitude is directly proportional to the optical pulse spectrum after spectral shaping $Y(t)$. As long as the time domain pulse input by the system is sufficiently narrow, the spectrum of the signal that is finally output after passing through the PD is largely determined by the frequency response characteristic of the optical pulse shaper and is directly in proportional. Hence, as long as the free spectral response of the spectral shaper is reasonably designed to vary linearly with the frequency and subsequently introduce the frequency–time mapping effect via the dispersive medium, the desired chirp signal can be generated.

## 3.4.2 LiDAR Data Processing Technology

### 3.4.2.1 Point Cloud Segmentation and Automatic Identification

After marking the point cloud data in the process of point cloud segmentation, the points in the spatial neighborhood with the same or similar attributes are classified into one class. In the past ten odd years, scholars have proposed a number of point cloud segmentation algorithms, which can be generally divided into direct segmentation and indirect segmentation. Many artefacts can be described by regular geometric shapes (such as planes, cylinders, and spheres), so the Hough transform can be used to extract geometric parameters directly from the point cloud data, and obtain geometric information of the object while implementing the segmentation. The indirect segmentation method uses progressive algorithms (such as cluster-based segmentation and region-growth algorithms) to compute spatial proximity and geometric derived values (such as local surface normal vectors and curvature).

Most of the central idea of segmentation algorithms based on local surface estimation classifies the unstructured point cloud data into spatially discrete patches with geometrical commonalities. However, as the transmitted laser pulses under this framework could not be properly described for many objects (such as grasses, trees, and leaves), it is necessary to study more general algorithms. Researchers in the field of machinery vision performs the spectral analysis by statistically analyze the dispersed scanned points in the local three-dimensional geometric features (points, lines, planes). For instance, under the supervised learning framework where the Gaussian Mixture Model is used to model for point cloud data recognition, and Bayesian classifier and graph cutting algorithms which are used to perform automatic point cloud data classification and recognition.

Based on the detection characteristics of the three-dimensional LiDAR, the point cloud density decreases from the center to the periphery in the ground projection, and the cloud density increases at the obstacle point, so the obstacle can be separated from the environment according to the point cloud density variation characteristics. The effect of laser point cloud segmentation is shown in Fig. 3.12.

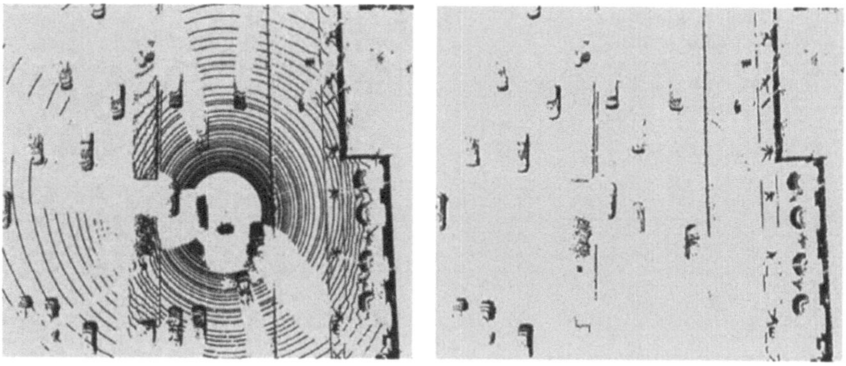

**Fig. 3.12** Laser point cloud segmentation. Left image: original point cloud, right image: segmented point cloud

1. **Hough Transform Method**

If a feature in the 3D space can be mathematically represented by a function F, such as a plane, a sphere, a cylinder, etc., the 3D-Hough transform can appropriately identify this feature.

Taking planar features as an example, in a three-dimensional Cartesian coordinate system, F is generally expressed as:

$$ax + by + cz + d = 0 \tag{3.17}$$

3D-HT uses the Eq. (3.18) to represent the plane features, which can convert the plane parameters into angle information.

$$\begin{aligned} &\rho = x\cos\theta\sin\varphi + y\sin\theta\sin\varphi + z\cos\varphi \\ &\theta \in [0^\circ, 360^\circ), \varphi \in [-90^\circ, 90^\circ] \end{aligned} \tag{3.18}$$

Wherein, the parameter $\theta$ represents the angle between the plane, **n**, normal to the *xoy* projection plane and the positive direction of the *x*-axis; $\varphi$ represents the angle between **n** and *xoy* plane; and $\rho$ represents the origin, O, distance from the plane, as shown in Fig. 3.13.

$n_\varphi$ represents the number of segments dividing the three-dimensional Hough space in the $\varphi$ direction, $n_\theta$ represents the number of segments divided in the $\theta$ direction, $n_\rho$ represents the number of segments divided in the $\rho$ direction, and the three-dimensional Hough space is divided into three blocks $n_\theta \times n_\varphi \times n_\rho$ in total which is used as a partition of the accumulator to perform the next step of voting. In fact, through the three-dimensional space point P, there is an infinite number of planes. However, due to the three-dimensional Hough discretization of the divided spaces, the number is limited to $n_\theta \times n_\varphi$ number of $\theta$ and $\varphi$ to determine the

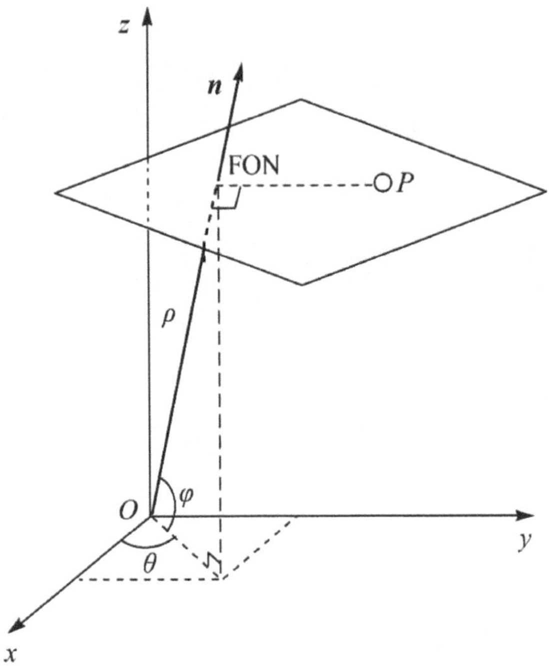


**Fig. 3.13** Hough transform in three-dimensional spatial plane

progeny of the Formula (3.18) ($\rho$ can also be determined similarly by $n_\theta \times n_\varphi$ number). This also mean that for every point P, there is a need to find $n_\theta \times n_\varphi$ number of planes that can satisfy the Formula (3.18) (by dividing the interval one by one to find), that is, each point P needs to vote in the three-dimensional accumulator $n_\theta \times n_\varphi$ times. Suppose that there are N number of points in the point cloud, then the whole process in the accumulator would require $N \times n_\theta \times n_\varphi$ times. Finally the accumulation result in the accumulator which displays the local peak point would indicate the plane that may exist.

2. **Clustering-Based Segmentation**

The spatial point cloud clustering problem can be defined as: giving a set of spatial points, finding a partitioning rule, and dividing the set of spatial points into subsets, so that the points in each subset are similar under the dividing rules.

For the clustering of arbitrary data sets, the common algorithms include: clustering based on partitioning, clustering based on hierarchy and clustering based on density. The methods based on partitioning comprises of K-means algorithm, K-medoid algorithm, X-means algorithm, etc. The hierarchy-based algorithms include the BIRCH algorithm, the CURE algorithm, etc.; the density-based method includes the DBSCAN algorithm. The following sections describe the K-means algorithm, the BIRCH algorithm, and the DBSCAN algorithm.

1. **K-Means Algorithm**

K-means algorithm is the most classical clustering method that is based on partitioning. The fundamental concept of the K-means algorithm is to initialize with K number of cluster centers randomly, based on the principle of the nearest sample point to be classified and assigned to each cluster. Then the centroid of each cluster is recalculated through averaging to determine the new cluster core. The computation is reiterated until the movement distance of the cluster core is less than the specified value.

Specifically, based on the K-means distance discriminating method, it will first focus on selecting k number of points randomly in space, and then use these k number of points as classification center points or average values to calculate the distance for each point to the selected k points, so as to classify each point to the k core cluster of the shortest or nearest distance. By traversing all the points, a completely new clustering result could be obtained. By reiterating the computation of the cluster centroids or average values, new k number of center points could be obtained. The computation is repeated until the clustering results no longer change or satisfy a specified criterion. For instance, one of the criteria could be to satisfy the squared difference requirement where the squared difference criterion is given as $\mathrm{E} = \sum_{i=1}^{n}\sum_{p=1}^{k}\left(K_p - M_i\right)^2$, when the distance sum of squares for all points satisfy the criterion of less than a set value, clustering stops. In this example, n is the number of all points, k is the number of classes, $K_p - M_i$ represents the distance from point i to the center of the class.

Through the k-means algorithm, we can input the number of classes and output a certain point that belongs to a certain class. The K-Means clustering algorithm is mainly divided into three steps:

(1) The first step is to find a clustering center for the points to be clustered;
(2) The second step is to calculate the distance from each point to the cluster center, and cluster each point into the cluster closest to the point;
(3) The third step is to calculate the coordinate average of all points in each cluster, and use this average as the new cluster center;

Repeat (2) and (3) until the cluster center no longer moves extensively or the number of clusters reaches the required value.

2. **BIRCH Algorithm**

The implementation of the B IRCH algorithm needs to be based on the clustering feature (CF) and the clustering feature tree (CF Tree), and hence these would be introduced first.

Clustering feature (CF): Given N data points $\{\overrightarrow{x_i}\}$, CF is defined as a triplet and is expressed as $\mathrm{CF} = \left(\mathrm{N}, \overrightarrow{LS}, \overrightarrow{SS}\right)$. Of which, N represents the number of data points; $\overrightarrow{LS}$ represents the linear sum of all the feature dimension vectors of all the data points in the CF representation; and $\overrightarrow{SS}$ represents the sum of the squares of all the feature dimension vectors of all the data points in the CF representation. CF also possesses additivity. Assuming, $CF_1 = \left(N_1, \overrightarrow{LS_1}, \overrightarrow{SS_1}\right)$ and $CF_2 = \left(N_2, \overrightarrow{LS_2}, \overrightarrow{SS_2}\right)$, then $CF_1 + CF_2 = \left(N_1 + N_2, \overrightarrow{LS_1} + \overrightarrow{LS_2}, \overrightarrow{SS_1} + \overrightarrow{SS_2}\right)$.

Clustering feature tree (CF Tree): CF tree is made up of CF which is used for storing cluster features by hierarchical clustering, as shown in Fig. 3.14. CF tree has three important parameters: the internal node balance factor B, balance factor leaf node L and the threshold value T. B defines the maximum number of off-spring of each non-leaf node; L defines the maximum number of sub-clusters per leaf node; T defines the maximum radius of the cluster, that is, all sample points in this CF must be with a radius of less than T.

The main process of the B IRCH algorithm is the process of building a CF tree. The process of inserting a new data point into a CF tree can be summarized as follows:

(1) From the root nodes in a top down manner, find the leaf node that is nearest to the new sample and the nearest CF node in the leaf node.
(2) If the distance between the new sample and the nearest CF node is less than the threshold T, a new sample is incorporated into the CF node, and the insertion ends. Otherwise go to step 3.
(3) If the number of CF nodes of the current leaf node is less than the threshold L, a new CF node is created and a new sample is inserted. A new CF node is then placed in the leaf node and all CF triplets on the path are updated, and the insertion ends. Otherwise go to step 4.

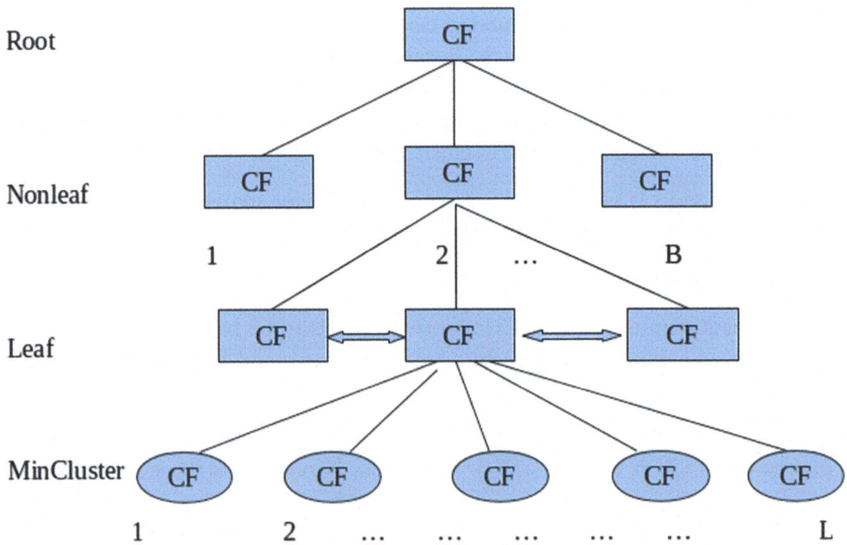


**Fig. 3.14** Clustering feature tree

(4) The current leaf node is divided into two new leaf nodes, and the two CF tuples with the furthest distance among all CF tuples in the old leaf node are selected and distributed as the first CF node of the two new leaf nodes. Place other tuples and new sample tuples into the corresponding leaf nodes according to the distance. In turn, check upwards if the parent node also needs to be split. If require, to split the same way as the leaf node.

In the final output of the CF tree, the sample points in each CF node in the tree is a cluster of clusters, and the clustering is then completed.

The main advantages of the BIRCH algorithm are: (1) Memory saving, all samples are on the disk, and the CF tree only stores the CF node and the corresponding pointer. (2) The clustering speed is fast. It is only necessary to scan the training set once to build the CF tree. The addition, deletion and modification of the CF tree are fast. (3) It is possible to identify noise points and data pre-processing could be conducted through preliminary classification.

The main disadvantages of the BIRCH algorithm are as follow: (1) Since the CF tree has a limit on the number of CFs per node, the result of the clustering may be different from the actual class distribution. (2) The data clustering result is not good on high dimensional features. Hence, MiniBatch K-Means can be adopted. (3) If the distribution cluster of the data set is not similar to a hypersphere, or is not convex, the clustering effect is not good.

3. **DBSCAN Algorithm**

DBSCAN is a density-based clustering algorithm. Such density clustering algorithms generally assume that categories can be determined by the closeness of the sample distribution. Samples of the same category are closely related to each other, that is, there must be samples of the same category that are located not too far from a particular sample.

(1) DBSCAN Density Definition

DBSCAN is based on a set of neighborhoods to describe the closeness of the sample set. The parameters ($\epsilon$, MinPts) are used to describe the closeness of the sample distribution in the neighborhood. Where $\epsilon$ describes the neighborhood distance threshold of a sample, MinPts describes the threshold of the number of samples in the neighborhood of a sample with a distance $\epsilon$.

Assuming the sample set is given as D $= (x_1, x_2, x_3 \ldots, x_m)$, then the specific density description of DBSCAN is defined as follows:

1. $\epsilon$-Neighborhood: for $x_j \in D$, the $\epsilon$-neighborhood consists of a set of samples D in the $x_j$ distance is not more than $\epsilon$ sub-sample set, i.e. $N_\epsilon(x_j) = \{x_j \in D | distance < (x_i, x_j) \leq \epsilon\}$, the number of the sub-sample set is referred to as $|N_\epsilon(x_j)|$;
2. Core object: for any one sample $x_j \in D$, if $\epsilon$-neighborhood corresponding $N_\epsilon(x_j)$ comprise of at least MinPts number of samples, and if $|N_\epsilon(x_j)| \geq$ MinPts, then $x_j$ is a core object.

3. Direct Density: if $x_i$ and $x_j$ are located in the $\epsilon$-neighborhood, and $x_j$ is a core object, then it is said that $x_i$ can be reached by the direct density of $x_j$. Note that the contrary is not necessarily true, i.e. it is incorrect to say that $x_j$ can be reached by the direct density of $x_i$, unless $x_i$ is also a core object.
4. Achievable Density: For $x_i$ and $x_j$, if sample sequence $p_1$, $p_2$,... $p_T$ exists, satisfies $p_1 = x_i$ and $p_T = x_j$, and $p_{t+1}$ is obtained directly by $p_t$ density, then we can say that $x_j$ is obtained directly by $x_i$ density. That is, the achievable density satisfies the transferability condition. The transferred samples in the sequence $p_1, p_2, \ldots, p_{T-1}$ are considered as the core object because only the core object could facilitate other sample densities to be derived directly. However, it is to note that the achieved density does not satisfy the symmetry condition. The achievable density could be derived from the asymmetry of the direct density.
5. Density Relatedness: For $x_i$ and $x_j$, if the core object sample $x_k$ exists, and results in $x_i$ and $x_j$ densities to be achieved through $x_k$ density, then $x_i$ and $x_j$ densities are considered to be related. Note that density relatedness satisfies the symmetry condition.

From Fig. 3.15, MinPts = 5, when the $\epsilon$-neighborhood of a particular sample comprises of at least 5 samples, a core object is formed. In the diagram, the red dots represent the core objects while the black dots represent non-core object. When the samples are located within the super sphere of a particular red denoted core object, the density could be obtained directly. The core objects which are connected by the green arrows in the diagram forms a sequence of samples for which density is reachable. All the samples within the $\epsilon$-neighborhood of the density-reachable sample sequence are each density connected.

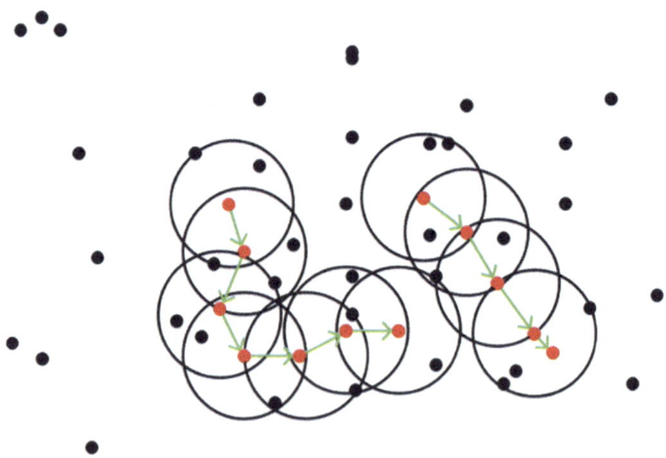

**Fig. 3.15** DBSCAN schematic diagram

(2) DBSCAN Density Clustering Algorithm

DBSCAN algorithm classifies the data points into three types: core points, i.e. within the half radius $\epsilon$ that comprises more than MinPts number of points; boundary points, i.e. within the half radius $\epsilon$ where the number of points are less than MinPts but are located within the neighborhood of the core point; and noise data, i.e. points which are neither core points nor boundary points.

The algorithm process can be expressed as:

(1) Select an unprocessed point, q, from the data set (not classified into any cluster or marked as noise) and examines within the neighborhood. If the number of objects contained within the neighborhood is not less than MinPts, a new cluster C is created where all the points are added to the candidate set N;
(2) For the unprocessed objects, q, within the candidate set N, examines the neighborhood. If it contains at least MinPts number of objects, add these objects into the candidate set N. If q has not been classified into any cluster, add q to C;
(3) Repeat step 2, continue checking for unprocessed objects within N until the current candidate set N is empty;
(4) Repeat steps 1 to 3 until all the objects are classified into a cluster or marked as noise. The clustering process is completed.

The main advantages of DBSCAN include: (1) Clustering of dense data sets of arbitrary shapes can be performed. In contrast, clustering algorithms such as the K-Means are generally only applicable to convex data sets. (2) It is possible to find abnormal points during the clustering process and is not sensitive to the abnormal points in the data set. (3) The clustering results are not biased. In contrast, the initial values of clustering algorithms such as K-Means have a great influence on the clustering results.

The main disadvantages of DBSCAN include: (1) If the density of the sample set is not uniform and the difference in cluster spacing is very different, the clustering quality is poor. At this time, DBSCAN clustering is generally not suitable. (2) If the sample set is large, the cluster convergence time will be longer. At this time, the KD tree or the ball tree which is established during the process of searching for the nearest neighbor could be scaled for improvement. (3) It is more complex to perform adjustment on traditional clustering algorithms like K-Means. Mainly, it requires the joint adjustment on the distance threshold, $\epsilon$, and the neighborhood sample size threshold, MinPts. The different parameter combinations have a greater effect on the final clustering influences (Fig. 3.16).

### 3.4.2.2 Point Cloud Filter

Pertaining to different types of point cloud data, different corresponding filtering method is required.

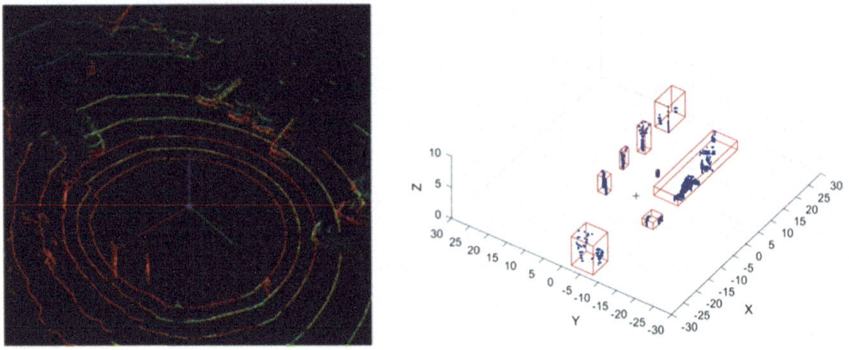


**Fig. 3.16** DBSCAN algorithm processing results (left image depicts original point cloud, right image depicts the clustering result)

1. **Airborne LiDAR Filtering**

The discrete LiDAR data point cloud is divided into ground points and non-ground point, which are then used in various fields, such as terrain applications, which is the main work of vehicle LiDAR data filtering. Common airborne LiDAR filtering methods include:

(a) **Interpolation based filtering**

The interpolation-based filtering include: (1) DEM interpolation; (2) TIN iterative cryptographic interpolation. The latter being the classical interpolation method.

Firstly, determine the boundary of the point cloud data. Through determining the top-left corner and lower right corner of the plane coordinates, the length, l, and width, w, of the data region could be confirmed. Based on parameter m (as shown in Fig. 3.17a), the plane is divided into a grid consisting of rows and columns. The number of rows and columns is calculated as shown in Eq. (3.19):

$$n_{row} = ceil\left(\frac{l}{m}\right), n_{column} = \mathrm{ceil}\left(\frac{\omega}{m}\right) \tag{3.19}$$

where $n_{row}$ is the number of rows, $n_{column}$ is the number of columns, and ceil (x) is a function that returns an integer value that is not less than x.

By selecting the lowest point in each grid and the four corner points as seed points, the elevation of the four corner points is equivalent to the elevation of the seed point that is nearest to the corner point. By establishing an initial triangulation, the point cloud is divided into seed point and to-be-determined point data.

TIN iterative encryption process is conducted by the continuous review of the points to be determined and updating the triangular nets, as shown in Fig. 3.17b, c.

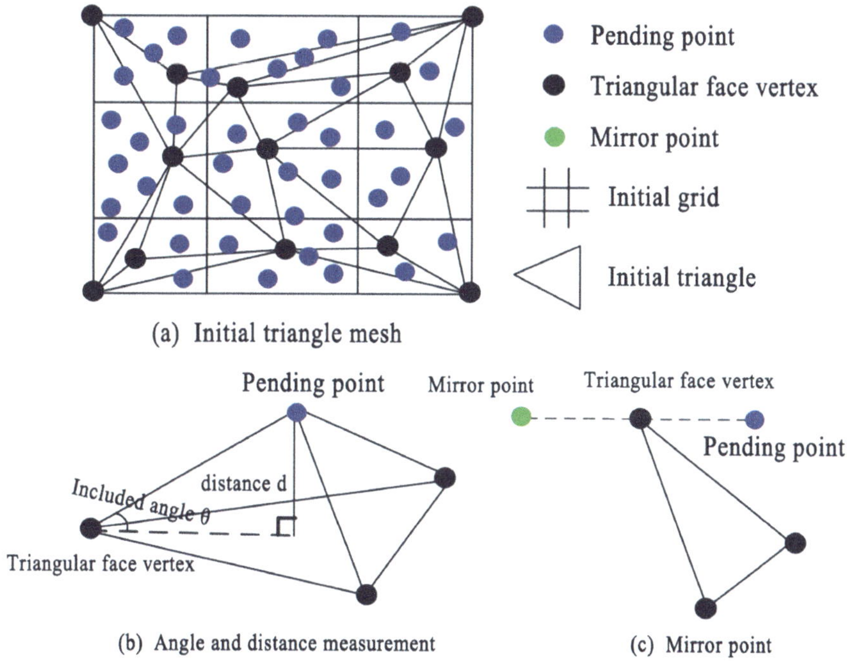


**Fig. 3.17** Description of TIN filtering process

(b) **Morphology**

By setting a high threshold value to separate ground points and non-ground points, the data is divided into a series of square cells, multiplying the size of the square unit in an iterative process.

(c) **Distance-Based Filtering**

The angle threshold is used to remove the side points, leaving behind the homogenous areas such as the ground points and the non-ground points at the top of the building, then perform clustering based on distance constraints, and finally counting the number of category data points. Obviously, most of the data points are ground points. Compared to the interpolation-based method, the cluster-based method is only applicable to flat regions. The algorithm is greatly affected by parameters, and especially sensitive to elevation threshold selection. Parameter automation cannot be achieved and excessive segmentation often occurs when methods similar to distance-constrained segmentation is employed. This resulted in a loss of detailed building information and is not suitable for complex interception of building treatment.

2. **Car LiDAR Filter**

The on-board laser LiDAR can acquire three-dimensional coordinate information like the airborne laser LiDAR, and the data processing must process massive point cloud data. However, the point density of the vehicle is many times higher than that of the airborne. If the frequency is the same and the scan point is set at 1 for a 1000 m high airborne scanner, the scan point for a 10 m distance car scanner would be 100. Hence, in theory, in a scenario where a car has passed by, any texture or minor undulation of a building could be reflected. Taking a house as an example, the car scanner scans the side of the house while the airborne scanner scans the roof top surface of the house. If the two are combined, a complete building model could be obtained. The car laser LiDAR is developed slightly later than the on-board LiDAR. Hence, many scholars borrowed the concept of separating ground and non-ground points when they were researching on the on-board LiDAR filtering, such as the method of scan line filtering.

Although the on-board laser scanning point cloud is spatially distributed as discrete point clouds, the point clouds located on different objects will exhibit different distribution characteristics in the same scan line profile. From the spatial distribution of the laser line-based scanning point cloud, there is usually no obvious change in the terrain or elevation for urban land, and hence, explains the ability to be approximated by light sliding in successive horizontal straight lines. The point cloud for the façade of the building can be approximated by light sliding in successive vertical lines. Hence, based on the scan characteristics of the building facade and the ground surface, the scan lines are segmented into different collective groups. The points in the collected group comprise mainly of the building elevation points and the ground points. The remaining point clouds are discrete points which are classified into other landmark points. Segmentation of scan lines means to process the laser generated points sequentially from left to right along each scan line. The different linear point sets are segmented based on the geometric relationships between adjacent laser points. Each segment belongs to a different linear target which will filter out the discrete points. The slope difference is used to detect the distribution relationship between adjacent points in the scan line profile. The phase interproximal point slope $\alpha_i$ is calculated using the formula below:

$$a_i = \arctan\frac{(Z_i - Z_{i-1})}{\sqrt{(X_i - X_{i-1})^2 + (Y_i - Y_{i-1})^2}} \tag{3.20}$$

The effect after filtering by the scan line is shown in Fig. 3.18.

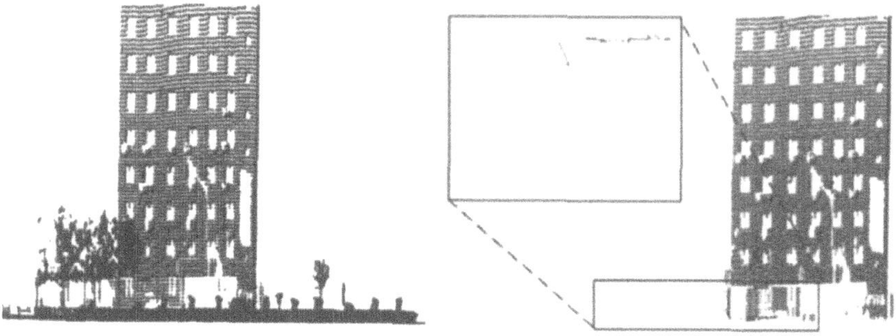

(a) Original Point Cloud (b) Scan Line-based Filtering Method

**Fig. 3.18** Comparison of before and after scan line filtering

# 3.5 LiDAR Application

## *3.5.1 Drone Application*

As the drone technology matures and costs gradually decline, the drone LiDAR technology starts to become popular. This means that the threshold of high-precision measurement and mapping technology is reduced, and hence it is no longer only limited to major, professional and well-funded projects by aircrafts. Although the cost of drone LiDAR is still much higher than photogrammetry, the acceptance of drone LiDAR technology has improved a lot compared to a few years ago.

UAV LiDAR technology has the characteristics of high automation, minimal weather impact, short data production cycle and high precision. It can be widely used in power line inspection, channel mapping, railway track and pipeline inspection, architectural and structural inspection, archaeology and culture, relics monitoring, three-dimensional measurement, forest resource survey, water conservancy, hydropower survey and design, tidal flat topography, crop growth monitoring, fluid mechanics modelling, digital elevation model and other fields. The application of the UAV laser radar will be described below by taking some of the fields as examples.

1. **UAV Autonomous Obstacle Avoidance System**

In lidar sensors, a single device integrates multiple independent components and will generate critical ranging data for precise positioning and safe navigation. The lidar sensor has an obstacle detection function in a wide range of fields of view, making it an ideal choice for UAV system sensing and evasion solutions.

The following picture in Fig. 3.19 shows the FLA (High Speed Light Unmanned Vehicle) UAV developed by the US Department of Defense under the DARPA (Defense Advanced Research Projects Agency) program, which specializes in the

**Fig. 3.19** FLA drone LiDAR for autonomous obstacle avoidance

development of high-tech for military applications. A laser radar is installed on the bottom of the UAV which can be used for autonomous obstacle avoidance. The drone can pass at a speed of 72 km per hour after recognizing the surrounding environment. After the obstacle ahead is detected, the obstacle can be advanced at a slower speed.

2. **Road Survey**

In conventional longitudinal cross-section measurement, usually spirit level, RTK, total station and theodolite devices are used to collect data. These methods are not only time consuming but also greatly limited by poor conditions of accessibility and unreachable places. The UAV laser radar measurement system can reduce the amount of manual work under the premise of ensuring accuracy, which is not only convenient but also safe. The UAV laser radar could be applied to the whole process of road survey, preliminary use of three-dimensional data and design requirements for road construction due to its promising operating performance. For mid-stage construction process, by using the three-dimensional laser radar data, the earthwork volume, road slope, road camber and other indicators measurement could be monitored. In the last stage, it could model a completed road in three-dimension to compare with the designed model for construction review and design deviation detection (Fig. 3.20).

3. **Power Inspection**

Due to the continuous expansion of the domestic power grid, long-distance transmission lines such as special (super) high-voltage lines are growing rapidly, and

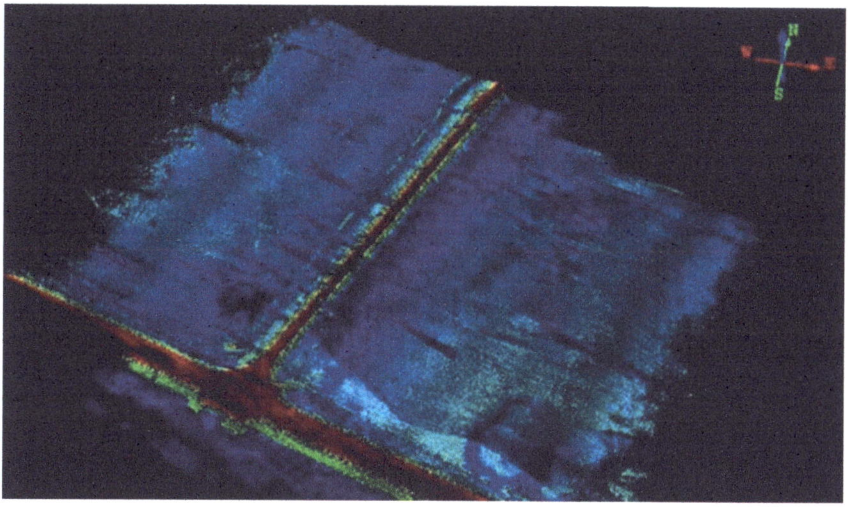

**Fig. 3.20** Application of UAV LiDAR measurement system in road survey

many transmission lines are being distributed in the mountainous areas. In order to maintain the daily power line and prevent the occurrence of electric accidents, it is necessary to conduct daily power lines inspection. At present, the main work of the power line inspection requires the manual patrol of the existing power lines by electricians. The work is tedious and laborious.

The use of drone laser radar technology for power line inspection is an advanced technical means. The use of this technology for power line inspection not only can greatly improve the work efficiency, but also greatly reduce the field work and reduce the cost of the line. The UAV laser radar technology can realize the real three-dimensional reconstruction of the electric pole tower and power line by collecting and processing the point cloud data of the electric power tower, power line, surrounding vegetation and other ground objects, thereby realizing the accurate measurement of vegetation and power lines and safety hazard detection which provides real and effective data to support decision making (Fig. 3.21).

### 3.5.2 Application in Unmanned Vehicle

Lidar plays the role of the "eye" in driverless cars, allowing you to quickly draw 3D panoramas based on scanned point cloud data. When the lidar finds an obstacle, it will control the vehicle to slow down or stop, and re-designate a safe route to proceed. The commonly use laser radar scan ranges up to a maximum of 200 m, the scanning frequency of up to 15 Hz, the angular resolution of 0.5°, the measurement resolution of within 0.1%, and accuracy within centimeters. Lidar can be divided

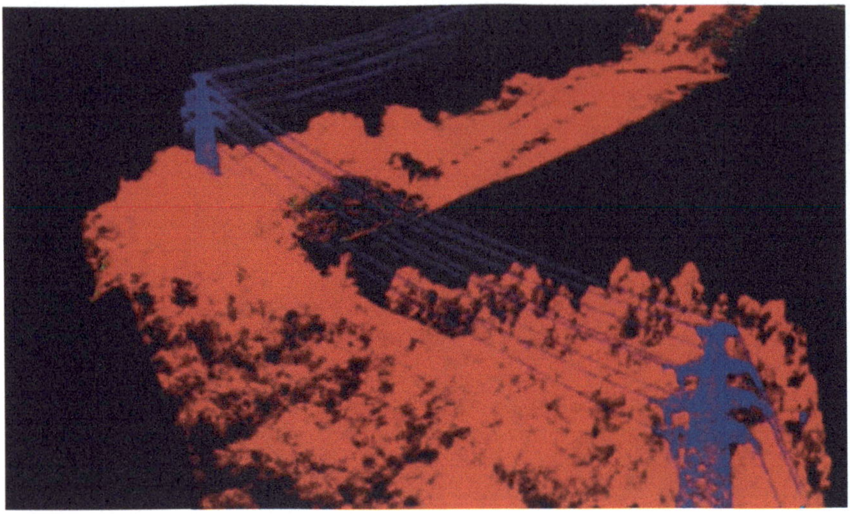

**Fig. 3.21** Application of UAV LiDAR measurement system during power inspection

into 1 line, 4 lines, 8 lines, 16 lines, 32 lines or 64 lines according to the laser beam (line). The more the laser line, the higher the angular resolution of the vertical direction and the laser point cloud density. An intelligent vehicle with a high degree of mobility takes on a longer measurement range and higher number of lines (8 lines, 16 lines or more). Hence, one of the major technical development in the application of LiDAR in autonomous vehicle is to increase the number of scan lines to ensure adequate detection range while hoping to reduce the weight by compressing the dimensions.

Lidar has two core roles in driverless operation:

1. **Identifying Surrounding Environment**

The 3D model of the surrounding environment of the car can be obtained by laser scanning, and the surrounding vehicles and pedestrians can be easily detected by using correlation algorithms to compare the changes between the previous frame and the subsequent frame environment. Four rotatable lidar sensors are mounted on the roof of the self-driving car, continuously emitting a weak laser beam to the surroundings, thereby realizing a 360-degree 3D streetscape around the car in real time. At the same time, combining with the 360-degree camera, can help the car to observe the surrounding environment. The system will further analyze the collected data to distinguish between consistent solid objects (lane dividers, exit ramps, park benches, etc.) and moving objects (pedestrians, oncoming vehicles, etc.). By aggregating all the data and adopting a corresponding algorithm to distinguish the surrounding environment, a corresponding response is made (Figs. 3.22 and 3.23).

**Fig. 3.22** Mounted position of LiDAR

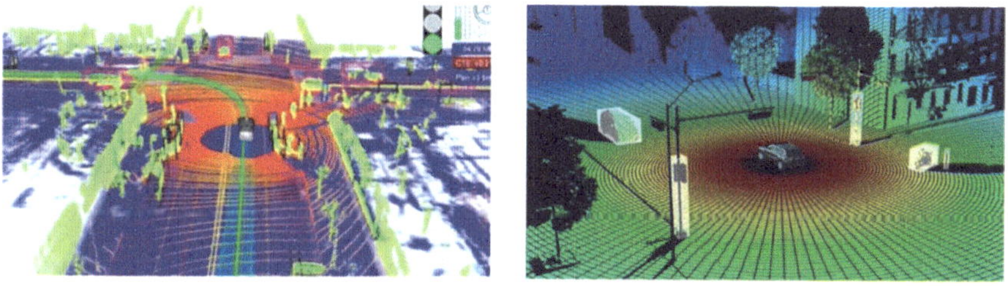

**Fig. 3.23** Illustration of LiDAR environment recognition

## 2. SLAM Strengthens Positioning

Another major feature of LiDAR is the Simultaneous Localization And Mapping (SLAM). The real-time global map can be compared with the features in high-precision maps to realize navigation and enhance vehicle positioning accuracy.

High-resolution maps can be obtained by combining LiDAR and the camera. A data collection vehicle is used to collect high-precision map data of HAD level. Two LiDARs on the roof (located at the rear) and four cameras (two at the front and two at the rear) are installed on the vehicle to acquire data with the required accuracy of 10 cm. This method enables the three-dimensional model building of road facilities such as road signs, road obstacles, lane markings, etc. This type of collected data is only used as a base. Road information is consistently being updated. With the increase in automated driving, there will be an increase in demand for real-time performance (Fig. 3.24).

**Fig. 3.24** High-resolution map of LiDAR

### *3.5.3 Application on Unmanned Surface Vehicle*

1. **Unmanned Surface Vehicle Obstacle Avoidance**

The unmanned surface vehicle (USV) is an intelligent small surface platform that can replace some of the surface boats to perform some cumbersome, complicated and even dangerous tasks in specific waters. If the mission is to be performed safely in a versatile and complex sea environment, the USV must be equipped with good situational awareness ability. In fact, the ability to perform accurate and stable obstacle detection is the required prerequisite and key technology for the USV.

The unmanned surface obstacle detection method can be divided into passive detection method and active detection method according to the different sensors used. Passive detection method refers to a method of sensing environmental information by using monocular vision or stereo vision sensor. Its main feature is that it can obtain rich colors display that uses an edge detection algorithm to extract the boundary between the sea and the sky. Thereafter, the feature of interest is being searched in the target area below the sky boundary line. However, it is often difficult to obtain accurate detailed information using the monocular vision detection method. Additionally, the image is highly vulnerable to changes in illumination. Furthermore, the stereoscopic detection method involves large amount of data which makes it harder to achieve the requirement of real-time processing. Active detection method refers to the use of ranging sensors such as LiDAR and marine radar to detect obstacles surrounding the boat. Marine radar is mainly used for target detection and tracking but usually it only provides position and velocity information of the target and is unable to identify the target's shape characteristics.

The radar also has a certain blind zone in close range. Due to LiDAR's high precision and strong anti-interference ability, it is very suitable for long-distance obstacle detection in unmanned surface craft. The unmanned surface boat generally uses a 360° omni-directional laser radar with an effective detection distance of 100 m and a vertical field of view of 30°. There are 16 scanning lines in the vertical direction and 300,000 laser point cloud data per second. In addition to obtaining the distance and orientation of the target, the reflection intensity of the target can also be obtained. The information plays an important role in the filtering of the clutter and the classification of the target (Fig. 3.25).

2. **USV Chart Mapping**

China has a sea area of 3 million square kilometers. Under such a vast sea area, nautical charts are vital to the navigation of ships and even to the national economy and national defense of the entire country. For the measurement of the chart, we usually conduct measurement through the sea survey ship. However, the sea survey ship has a deep draught as we can witness in sea survey ships such as the sea patrol 01, the sea patrol 166, Li Siguang and Qian Xuesen survey ships. There are many island reefs in China's more than 3 million square kilometers of sea area. It is estimated that there are more than 3500 in the East China Sea and more than 1700 in the South China Sea. With hundreds of thousands of square kilometers of seas less than 5 m deep, mapping sea maps around these islands and reefs is obviously not possible. Hence, USV can be deployed to solve this problem. The unmanned boat has a shallow draught, and can achieve self-propelled navigation, autonomous obstacle avoidance, and independent mapping of islands and reefs. The unmanned boat is used to detect and map the shallow depth of the sea area so that the sea vessel can avoid these dangerous areas.

In a laser-scanning underwater imaging system, a laser is irradiated with a continuous or high repetition rate of laser light, and the reflected signal is collected

**Fig. 3.25** LiDAR application on USV

by the detector. The laser beam performs line by line scanning on the target to obtain high-resolution imaging. Ideally, the diffraction-limited 532 nm laser beam can maintain a 10 mm diameter at 50 m range. When the working conditions are as follows: the output is 532 nm generated by frequency doubling; the laser pulse energy is 100 uJ; the repetition frequency is 5 kHz; and the pulse width is 10 ns, the laser underwater imaging system can detect depth in the ocean water to about 50 m, while in the turbid sea quality it can also detect a depth to about 15 m. This depth is sufficient to satisfy the draft of the sea survey vessel. As a result, these dangerous areas can be effectively avoided.

Previously when the Xuelong expedition ship was out on an expedition in the Ross Sea Antarctica, due to the lack of the surrounding sea map information, the ship was unable to anchor and stop the engine which resulted in a great cost. After using the USV for chart mapping, better anchor points could be found.

It is well known that the East China Sea has the greatest number of island reefs. In the past, the sea map data of the surroundings of the island reefs are obtained through satellite remote sensing. The data obtained from satellite remote sensing has especially high data error which can cause many difficulties for sea navigation. Therefore, some islands and reefs in the East China Sea can make use of the USV to conduct mapping.

From the diagram in Fig. 3.26, the blue colored indicates that the water is relatively deep. This is the data obtained by the satellite remote sensing. Based on this data, it shows that a ship could pass through the waters. However, after the USV is deployed to map the surroundings, it shows the entire zone is red colored. The redder the zone, the shallower is the waters. The shallowest areas indicated of a depth of only 0.85 m. If the data from the satellite remote sensor had been used for navigation, it could have brought upon great danger. Hence, through the use of the

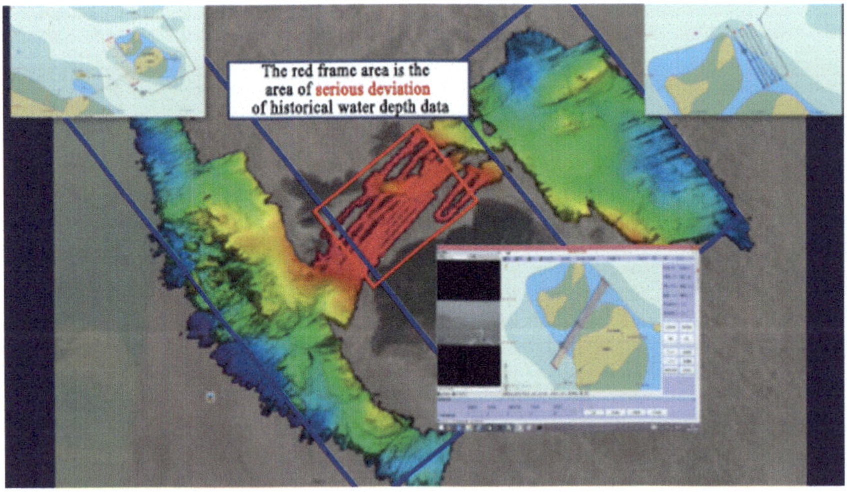


**Fig. 3.26** Mapping of the rhubarb reef

USV chart mapping could supplement the data obtained around the island reefs and generate a three-dimensional water depth map which can promote the ocean safety.

## Bibliography


1. Xu H (2014) Research on point cloud segmentation based on clustering method. Harbin Institute of Technology
2. Yang X (2005) Research on several key technologies in cluster analysis. Zhejiang University
3. Li M, Li G, Wang L et al (2015) 3D hough transform application in laser point cloud feature extraction. Surveying Mapp 2:29–33
4. Yuan X, Zhao C (2011) A laser point cloud clustering algorithm for robot navigation. Robot 33(1):90–96
5. Zhang M, Liu X, Fu R et al (2017) A laser point cloud clustering algorithm for road obstacle recognition. Laser Infrared 9:1186–1192
6. Zhang H, Chen J, Wei H (2015) An improved BRICH algorithm and its application. Softw Guide 14(10):45–47
7. https://www.cnblogs.com/pinard/p/6179132.html
8. https://www.cnblogs.com/pinard/p/6208966.html
9. https://baike.baidu.com/item/DBSCAN/4864716?fr=aladdin
10. Zou J (2017) Research on optical generation and pulse compression technology of linear frequency modulation signal. University of Electronic Science and Technology of China
11. Klauder JR, Price AC, Darlington S et al (1960) The theory and design of chirp radars. Bell Syst Tech J 39(4):745–808
12. Wang R (2004) Radar pulse compression technology and its time-frequency analysis. Northwestern Polytechnical University
13. Richards MA (2005) Fundamentals of radar signal processing. McGrawHill, New York
14. Li Z (2012) Research on several key issues in microwave photon signal processing. Zhejiang University
15. Zhu J (2011) Radar signal analysis and simulation implementation. Beijing Jiaotong University
16. Caputo M, Denker K, Franz MO et al (2018) Support vector machines for classification of geometric primitives in point clouds
17. Kang Z, Yang J, Zhong R (2016) A Bayesian-network-based classification method integrating airborne LiDAR data with optical images. IEEE J Sel Top Appl Earth Obs Remote Sens 99:1–11
18. Fang H, Lin X, Duan M et al (2015) Object-oriented vehicle LiDAR point cloud filtering method. J Survey Mapp 40(4)
19. Gao E, Zheng Y (2012) Overview of point cloud data filtering methods. Sci Technol Inf 33:4
20. Zeng N (2018) Overview of point cloud data filtering and classification methods. Urban Geogr (2)
21. Yang Y, Zhang Y, Ma Y et al (2010) Point cloud filtering method of vehicle laser radar based on scanning line. J Survey Mapp Sci 27(3):209–212

# Chapter 4
# Machine Vision

## 4.1 Introduction

Machine vision means to endow the machine with visual perception system which provides similar biological visual ability to facilitate situational awareness. The machine vision system converts the ingested object into an image signal by an image capturing device, transmits it to a dedicated image processing system and converts it into a digital signal according to the pixel distribution, width, color, etc. The image system performs various computations on the signals to extract the target characteristics. By discerning the results, the site operations are being controlled. The main research goal of machine vision is to enable computers to have the ability to recognize three-dimensional environmental information through two-dimensional images, and to sense and process geometric information such as shape, position, posture, and motion of the objects in a three-dimensional environment.

As a specific field of optoelectronic technology application, machine vision is widely used in many industries such as in microelectronics, electronic products, automotive, medical, printing, packaging, scientific research, military, etc. It is highly impactful and plays a very important role in image-based intelligent data acquisition and processing. Visual technology can effectively deal with the detection and recognition of specific target objects (such as face, handwritten characters or goods), image classification and subjective image quality assessment. In the recent years, the rapid development of deep learning has not only resolved many difficult visual problems, but also improved the quality of image recognition and accelerated the progress of the computer vision and the artificial intelligence related technologies. More importantly, it changed our traditional way of processing visual problems. At present, the visual technology has a huge commercial market in internet applications such as image search, product recommendation, user behavior analysis, and face recognition. At the same time, it has vast application potential in



X. Bi, *Environmental Perception Technology for Unmanned Systems*, Unmanned System Technologies, https://doi.org/10.1007/978-981-15-8093-2_4

advance technology industries such as intelligent robots, unmanned autopilots and drones, as well as, in scientific fields such as in biology, medicine and geology.

## 4.2 Machine Vision Concepts and Characteristics

### *4.2.1 Basic Concepts of Machine Vision*

The terms 'computer vision' and 'machine vision' are sometimes indistinguishable as it can be witnessed in many documents. In fact, the two terms are considered distinctive and yet related. Computer vision adopts a combination of image processing, pattern recognition and artificial intelligence technology, focusing on the calculation and analysis of one or more images. The image can be acquired by a single or multiple sensor, or it can be a sequence of images acquire by a single sensor at different times. The analysis is to facilitate the recognition of target object by determining the position and posture of the target object so as to decipher the three-dimensional environment. In computer vision research, geometric model, complex knowledge representation, use of model-based match and search techniques, bottom-up search strategies, top-down hierarchical and heuristic control strategies are commonly adopted.

Machine vision, on the other hand, focuses on the technical engineering of computer vision, and can automatically acquire and analyze specific scenes to control the corresponding behavior. Specifically, computer vision provides the theoretical and algorithmic basis for image and scene analysis for machine vision. In return, machine vision provides sensor model, system construction and implementation means for computer implementation. Therefore, it can be considered that a machine vision system is one that can automatically acquire one or multiple images of the target object which then processes, analyzes and measures the various extracted features of the image(s) to quantitatively analyze and interpret the results. Thereby, acquire certain understanding about the target object to make a corresponding decision. The functions of the machine vision system include: object positioning, feature detection, defect determination, target recognition, counting and motion tracking.

Machine vision systems are typically computer centric. It is mainly composed of modules such as the visual sensor, high-speed image acquisition system and a dedicated image processing system, as shown in Fig. 4.1.

Firstly, the image is taken by the camera which is then analyzed by the control unit. A two-dimensional structure of a semiconductor detector is mostly used as a radiation sensor in the visible and near-infrared spectral ranges. An internal spot effect is used at this point, where an incoming photon causes one of the electrons in the valence band to enter the conductive band.

In a CCD sensor (charge coupled device), the free-band particles thus generated are collected in a potential well and then read out line by line in accordance with the

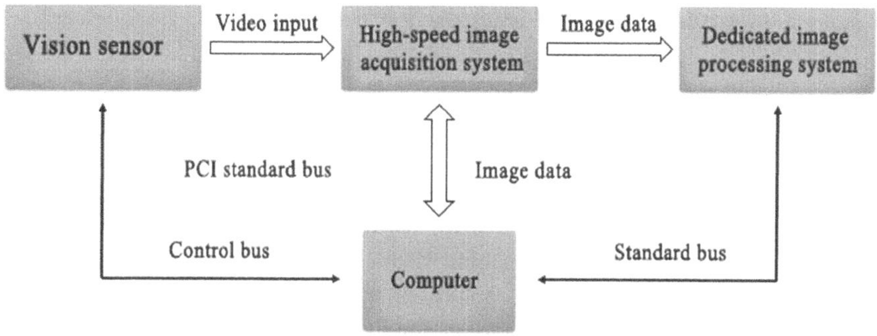


**Fig. 4.1** Basic components of machine vision system

bucket queue delay line principle, thereby producing an image signal proportional to the light efficiency. Although CCD technology has advantages in its high light sensitivity, light collectors are increasingly implemented by CMOS (Complementary Metal Oxide Semiconductor) technology. CMOS technology is currently the standard technology for integrating logic circuits with significantly lower current consumption and less responsive cooling than thermal noise. Therefore, within the scope of this technology, it is envisaged that the camera or even the entire image sensor will only have one uniquely integrated conversion circuit (system chip). The CMOS sensor allows for the optional selection of readout pixels. For example, for highly dynamic applications, the area of the associated image content can be read with very high read rates up to the MHz range. Therefore, the CMOS circuit allows a relatively simple realization of the nonlinear characteristic curve between the incident light efficiency E and the obtained image signal U. Similar to the human eye, the logarithmic perceptual light stimulation according to Weber-Feller's law is given as:

$$U = c \ln \frac{E}{E_0} \tag{4.1}$$

In the formula, $c$ and $E_0$ are constant.

In addition, it is also possible to implement the radiation sensor in a logarithmic manner, whereby the tens of powers of the luminance difference and the constant contrast resolution of the same image, $\Delta E/E$, can be differentiated. Due to the high data transfer rate of the image sensor, the image analysis control unit not only adopts the freely-programmable logic processor but also uses a digitalized component or a digital signal processor (the DSP dedicated component).

### 4.2.2 Characteristics of Machine Vision

Machine vision is a comprehensive technology that includes image processing, mechanical engineering, control, electric light source illumination, optical imaging, sensors, analog and digital video technology, computer hardware and software technology (image enhancement and analysis algorithms, image cards, I/O cards, etc.), and human-machine interface technologies, etc. These technologies can be coordinated to form a complete machine vision system. A typical machine vision application system includes image capture, light source system, image digitization module, digital image processing module, intelligent judgment decision module and mechanical control execution module.

The most basic feature of the machine vision systems is the increased flexibility and automation of the work. Machine vision is often used over artificial vision in hazardous work environments that are not suitable for manual work, where artificial vision is difficult to satisfy the requirements and where automation is needed. At the same time, in the process of large-scale repetitive industrial production, the machine vision inspection method can greatly improve the efficiency and automation of production. Machine vision emphasizes on the practicality, the ability to adapt to a variety of different environments, and the need to have reasonable cost performance, universal communication interface, high fault tolerance and security, strong versatility and portability. It emphasizes real-time performance and requires high speed and precision. Machine vision is called the eye of the automation and has a wide range of applications in the fields of national economy, scientific research and national defense construction.

With the help of the machine vision systems and sensor systems, the principle of perception that is the closest to the visual information adopted by people can be realized. Benefiting from the consistently decreasing prices of the cameras and analytical devices, image sensors are beginning to be used in more applications. One of the decisive advantages of video sensors is that they provide the most comprehensive display of the information compared to other environmental sensors. At the same time, the analysis of comprehensive image information poses a huge challenge to signal processing.

According to the latest state-of-the-art technology standard, a general comparison between the video sensor and other sensor technologies is conducted. The results are as shown in the table below. The camera's basic measurement includes only the brightness mode which can be converted into 3D information only by a suitable image analysis program. Lidar or long-range imaging systems provide a direct application for vehicle guidance with their direct 3D position measurement. Radar sensors can even directly make use of the Doppler effect to measure the radial velocity. Nonetheless, the vision sensor possesses great potential in providing more abstract information in comparison to the general environmental sensors. Given its proximity to human perception, the image sensor's machine vision system can achieve a high degree of transparency in its functions.

| Typical parameters, characteristics and potential | Vision | LiDAR | Millimeter-wave radar | Long-range imaging system |
|---|---|---|---|---|
| Wavelength/m | $10^{-7} \sim 10^{-6}$ | $10^{-6}$ | $10^{-3} \sim 10^{-2}$ | $10^{-7} \sim 10^{-6}$ |
| Weather condition dependence | Yes | Yes | Low | High |
| Resolution (# measured value) | | | | |
| Level | $10^2 \sim 10^3$ | $10^2 \sim 10^3$ | $10^1 \sim 10^2$ | $10^1 \sim 10^2$ |
| Vertical | $10^2 \sim 10^3$ | $10^1 \sim 10^2$ | $10^1$ | $10^1 \sim 10^2$ |
| Time | $10^1 \sim 10^5$ | $10^1$ | $10^1$ | $10^1$ |
| Main measurement | | | | |
| Position | Weak | Strong | Strong | Strong |
| Speed | Weak | Weak | Strong | Weak |
| Brightness mode | Strong | Strong | Weak | Strong |
| Function example | | | | |
| Object measurement | Strong | Strong | Strong | Strong |
| Object recognition | Strong | Strong | Medium | Medium |
| Lane/traffic sign recognition | Strong | Medium | Weak | Weak |

## 4.3 Camera Classification and Principle

### *4.3.1 Camera Components*

The camera is mainly composed of a lens, a photosensitive element, a processing circuit, and a power source.

(1) Lens

The basic function of the lens is to achieve the beam transformation (modulation). In machine vision systems, the main function of the lens is to project the image of the target on the photosensitive surface of the image sensor. The quality of the lens directly affects the overall performance of the machine vision system. The proper selection and installation of the lens is an important part of the machine vision system design (Fig. 4.2).

(2) Photosensitive Element

The photosensitive element is an image sensor (optical chip and photosensitive chip). It is a geometric silicon crystal piece with tens of thousands of photosensitive dots (pixels) that can sense and record information about external light and convert it into a current, which is then converted into digital information. The

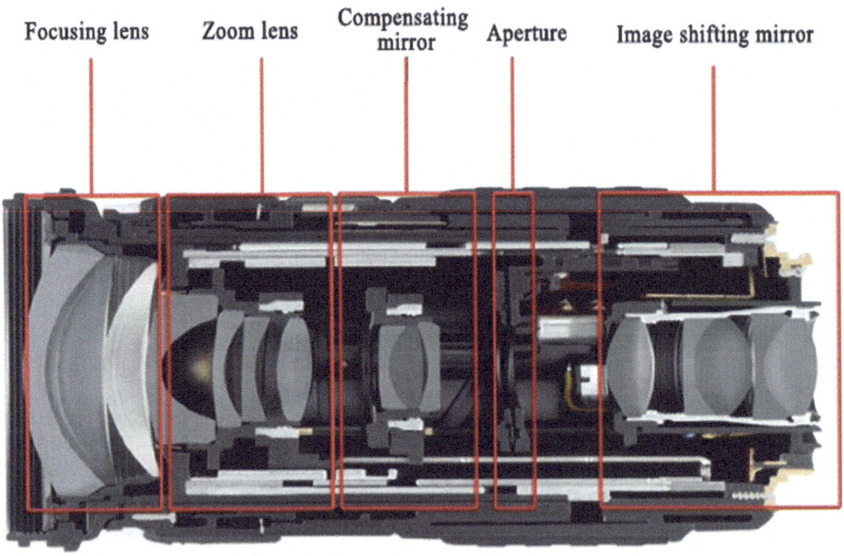


**Fig. 4.2** Camera control mechanism

photosensitive element can be classified into two types: a CCD (Charge Coupled Device) and a CMOS (Complementary Metal-oxide-Semiconductor).

(3) Processing Circuit

The processing circuit converts the digital information obtained by the photosensitive element into a data file of the digital image.

## *4.3.2 Classification of Cameras*

According to the camera function, we can classify the cameras into three major types, namely, monocular, binocular (stereo) and deep camera (RGB-D) as shown in Fig. 4.3. Intuitively, the monocular camera has only one camera and the binocular camera has two. The RGB-D principle is more complicated which usually involves multiple cameras. Other than its ability to capture color images, it can also read each pixel distance from the camera.

### 4.3.2.1 Monocular Camera Model

The imaging model of a monocular camera can be likened to using a pinhole camera model which makes use of the linear transmission of light that projects a three-dimensional object onto a two-dimensional imaging plane. The main

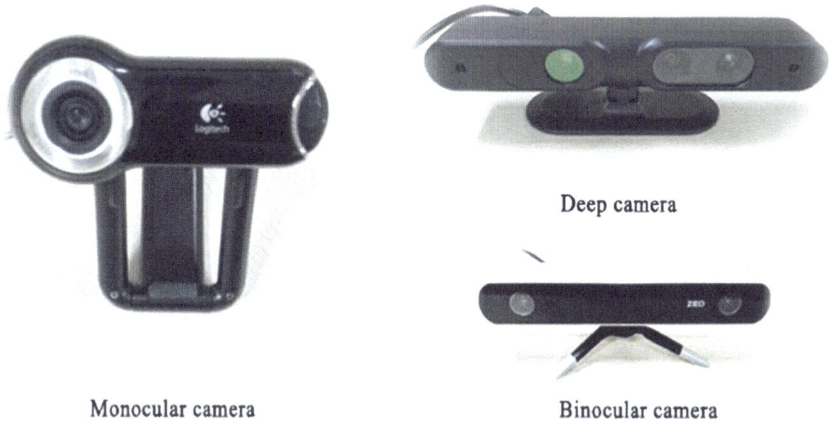


**Fig. 4.3** Classification of commonly used cameras

components of the pinhole camera model comprise of: the optical center (projection center), the image plane, the optical axis, etc., as shown in Fig. 4.4.

In the camera coordinate system, a point, $P$, in real space goes through the orifice, $O$, to project an image point, $P'$, on the physical imaging plane. The coordinates of the point, $P$, in the camera coordinate system are set as $(X, Y, Z)$ and the coordinates of the image point, $P'$, in the physical imaging coordinate system

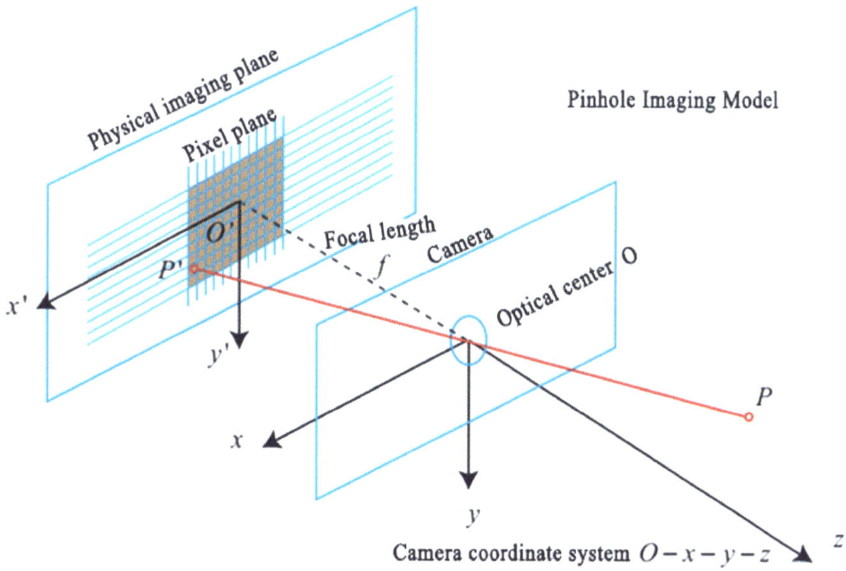


**Fig. 4.4** Pinhole imaging model

are set as $(X', Y')$. The coordinates in the image pixel coordinate system are set as $(u, v)$. Additionally, the distance from the physical imaging plane to the orifice is set as $f$ (focal length). According to the similarity of the triangle, the following is given:

$$\frac{Z}{f} = -\frac{X}{X'} = -\frac{Y}{Y'} \tag{4.2}$$

In fact, the camera does not output an inverted image. The camera would flip the image which is equivalent to performing a symmetry of the physical imaging plane to the front of the camera, and hence, Eq. (4.2) can be re-written as (Fig. 4.5):

$$\frac{Z}{f} = \frac{X}{X'} = \frac{Y}{Y'} \tag{4.3}$$

Wherein, between the pixel coordinate system and the physical imaging plane coordinate system, a zoom and an origin shift are distinguished, and the relationship can be expressed by the Formula (4.4):

$$\begin{cases} u = \alpha X' + c_x \\ v = \beta Y' + c_y \end{cases} \tag{4.4}$$

Let $f_x = \alpha f, f_y = \beta f$, the following relationship can be obtained:

$$\begin{cases} u = f_x \dfrac{X}{Z} + c_x \\ v = f_y \dfrac{Y}{Z} + c_y \end{cases} \tag{4.5}$$

Using the homogenous coordinates, the relationship can be represented in matrix form:

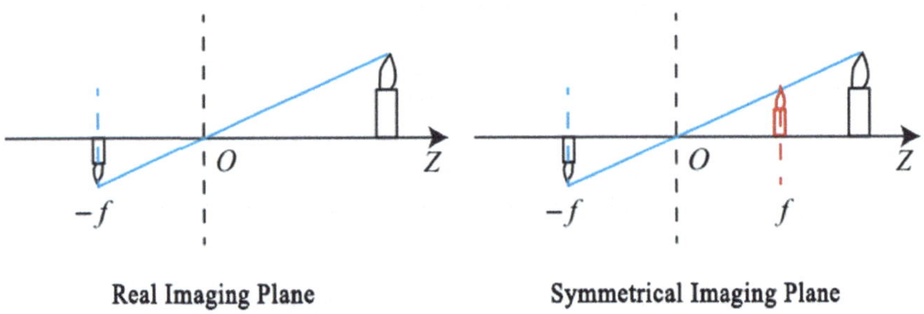


**Fig. 4.5** Diagrams illustrating the real imaging plane and symmetrical imaging plane

$$Z\begin{bmatrix} u \\ v \\ 1 \end{bmatrix} = \begin{bmatrix} f_x & 0 & c_x \\ 0 & f_y & c_y \\ 0 & 0 & 1 \end{bmatrix}\begin{bmatrix} X \\ Y \\ Z \end{bmatrix} = \mathbf{K}\begin{bmatrix} X \\ Y \\ Z \end{bmatrix} \tag{4.6}$$

In Eq. (4.6), the matrix, **K**, which is composed of intermediate quantities is known as the internal parameter matrix of the camera. It can generally be considered that the internal parameters of the camera are fixed after leaving the factory and will not change during use.

#### 4.3.2.2 Binocular Camera Model

A binocular camera generally consists of two horizontally placed cameras which forms the left eye and the right eye. It can also be configured to consist two vertically placed cameras. However, the mainstream binoculars are generally made in the horizontal manner, in which, we can treat both cameras as pinhole cameras. Since the cameras are placed horizontally, the center of the aperture of both cameras is positioned on the x-axis. Their distance is known as the binocular camera baseline (baseline is denoted as $b$) which is an important parameter in the binocular camera.

Now, consider a spatial point, $P$, that forms an image separately in the left and right eye and is marked as $P_L, P_R$ respectively. These two imaging positions are different due to the presence of the camera baseline. Ideally, since there is only displacement along the x-axis for both the left and right cameras, the image, $P$, will be different only on the x-axis (corresponding to the $u$-axis of the image). Hence, the coordinates on the left side are set as $u_L$ and the right side as $u_R$. Therefore, their geometrical relationship is as shown on the right side of Fig. 4.6. According to the similarity of the triangles $P - P_L - P_R$ and $P - O_L - O_R$, the equation is obtained:

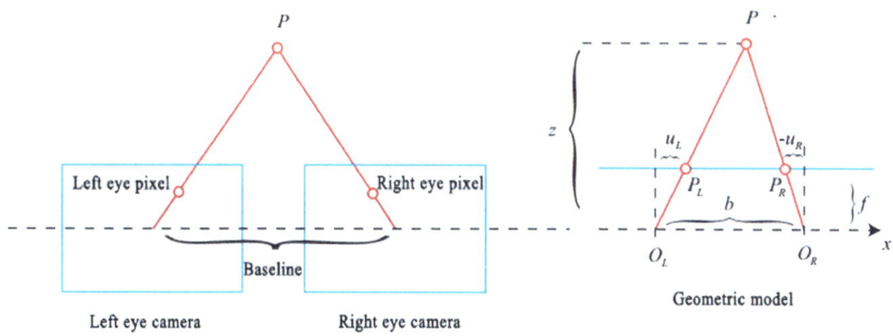


**Fig. 4.6** Imaging model of a binocular camera ($O_L, O_R$ are the left and the right center of aperture respectively; the blue box is the imaging plane; $f$ is the focal length; $u_L$ and $u_R$ are the coordinates of the imaging plane wherein $u_R$ is negative and hence the distance marked in the diagram is indicated as $-u_R$)

$$\frac{z-f}{z}=\frac{b-u_L+u_R}{b} \tag{4.7}$$

Which could be expressed as:

$$z=\frac{fb}{d},d=u_L=u_R \tag{4.8}$$

Wherein $d$ is the difference between the abscissa of the left and right image, known as disparity. Based on parallax, we can estimate the distance of a pixel from the camera.

#### 4.3.2.3 RGB-D Camera Model

The components of the RGB-D camera include at least one transmitter and one receiver in addition to the normal camera. Compared to the binocular cameras that calculate depth by parallax, RGB-D cameras can actively measure the depth of each pixel. At present, RGB-D cameras can be divided into two categories according to the principle:

1. Through Infrared Structured Light to measure pixel distance; and
2. Through Time-of-flight (ToF) to measure pixel distance (Fig. 4.7).

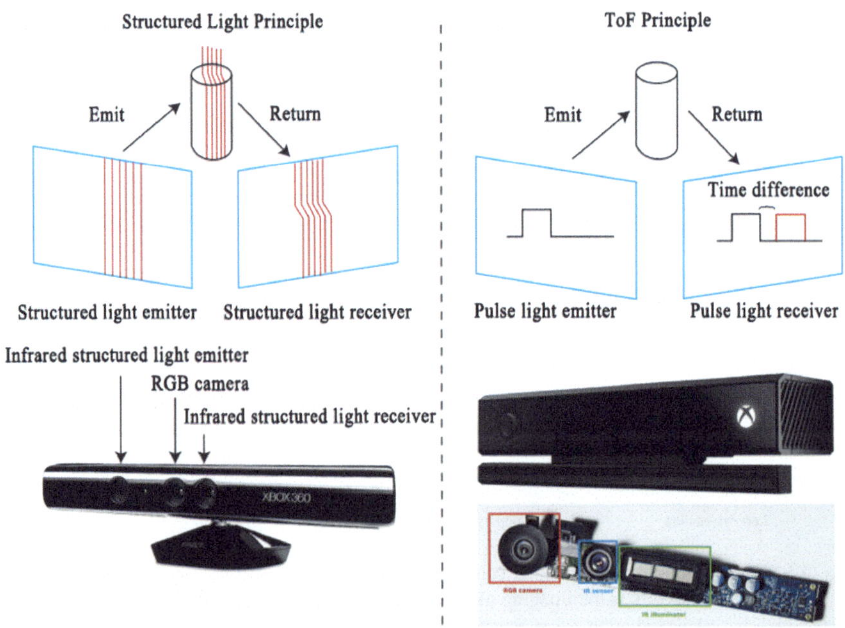


**Fig. 4.7** Schematic diagram of the RGB-D camera

RGB-D cameras need to emit a beam of light (usually infrared light) to the target regardless of structured light or ToF. According to the structured light principle, the camera calculates the distance of the object from itself based on the returned structured light pattern. On the other hand, in ToF, the camera emits pulsed light to the target and then determines the distance of the object from itself based on the time of flight of the beam between the transmission and the return. The ToF principle is very similar to a laser sensor. However, the laser acquires the distance through point-by-point scanning while the ToF camera can obtain the pixel depth of the entire image which displays the characteristic of the RGB-D camera.

After measuring the depth, the RGB-D camera usually completes the pairing between the depth and the color map pixels according to the position of each camera at the time of production, and outputs a corresponding color map and depth map. At the same image position, we are able to read the color and distance information, calculate the 3D camera coordinate of the pixel and generate point cloud. The RGB-D camera can measure the distance of each pixel in real time. However, due to this type of emission–acceptance measurement, its range of usage is limited. RGB-D cameras that use infrared for depth measurement are susceptible to interference from sunlight or the infrared transmitted by other sensors. Hence, it is not suitable for outdoor usage. Concurrent use of multiple cameras would also cause interference with one another. For objects with a transmissive material, the position of these points cannot be measured because the reflected light is not received. In addition, RGB-D cameras have some disadvantages in terms of cost and power consumption.

## 4.4 Machine Vision Technology

### *4.4.1 Traditional Machine Vision Technology*

Machine vision is a science that studies how to enable a machine to "see", that is, the use of a camera and a computer instead of the human eye to identify, track, and measure a target.

The main task of the machine vision is to process the collected images or videos to obtain the corresponding information of a scene. The main tasks of machine vision are as follows.

#### 4.4.1.1 Object Detection

Object detection is the first step in visual perception and an important branch of computer vision. The object detection makes use of a frame to mark the position of the target and then assign a category to the object as shown in Fig. 4.8.

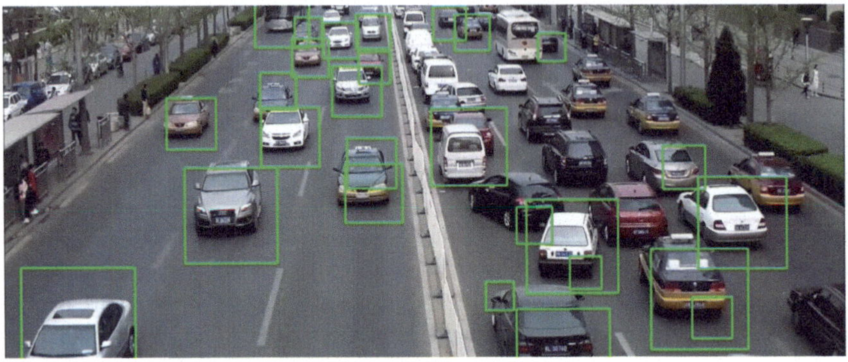

**Fig. 4.8** Vehicle detection result

With the development of science and technology, there are increasingly more methods for object detection. The conventional classical algorithms mainly include: the combination of the histogram of oriented gradient and the support vector machine algorithm, DPM algorithm and the combination of Haar feature and Adaboost algorithm. The combination of the histogram of oriented gradient and the support vector machine can achieve better results in the unmanned system. Hence, this section shall mainly introduce this method.

1. Histogram of Gradient Direction

Histogram of Oriented Gradient (HOG) feature is a descriptor used for object detection. The HOG feature is obtained by calculating and counting the histogram of oriented gradient of a localized region of the image.

The idea of HOG is not to use each individual gradient direction of each individual pixel of the image, but to group the pixels into small units. Pertaining to each cell, all the gradient directions are being calculated and then categorized into multiple direction regions, summarizing the magnitude of the gradient in each sample. Therefore, a stronger gradient contributes more weight to its case and the effect of small random orientation due to noise is reduced. The histogram provides an image of the dominant direction of the cell. By performing the same on all the cells provides a representation of the image structure. HOG can maintain good invariance to image geometric and optical deformation, and the shape of the detected image can also have some subtle changes. These subtle changes can be ignored without affecting the detection effect.

**Specific Operation of the HOG Feature**

(1) Normalization

In order to reduce the impact of lighting factors, the entire image needs to be normalized first. In the texture intensity of the image, the local surface exposure contribution is of relatively large proportion, so this compression process can effectively reduce the shadow and illumination variations of the image. As the color information is not effective, it is usually first converted to a grayscale image.

(2) Image Gradient Calculation

Calculatc thc gradicnt of the horizontal and vertical directions of the image, and then calculate the gradient direction value of each pixel position accordingly. The derivation operation can not only capture the contour and some texture information, but also further weaken the influence of illumination. The effect of the gradient calculation is shown in Fig. 4.9.

(3) Construction Direction Histogram

Each pixel in a cell unit votes for a direction-based histogram channel. The weighted voting is adopted, that is, each vote is weighted, and this weight is calculated based on the gradient of the pixel. The weight itself or its function can be used to represent this weight. The actual test shows that using the magnitude to represent the weight is able to achieve the best effect. Notwithstanding the above, the function of the magnitude could also be selected to represent the weight such as the square root of the amplitude, the square of the amplitude and the truncated form of the amplitude, etc.

(4) Combining Cell Units into Large Intervals

By combining the individual cell units into large spatially connected intervals, the HOG descriptor becomes a vector consisting of histogram components of all cell units in each interval. These intervals are mutual weight stack which means the output of each unit cell are repeatedly applied to the final descriptor (Figs. 4.10 and 4.11).

2. Support Vector Machines

Support Vector Machine (SVM) is a common identification method. In the field of machine learning, it is a supervised learning model that is commonly used for pattern recognition, classification, and regression analysis.

SVM is a powerful learning machine. Its fundamental theory is based on statistical learning theory such as the statistical VC theory and the structural risk minimization principle. Statistical learning theory is the theory of studying small-scale statistics and prediction. The support vector machine is developed from the optimal classification extension in the case of linear separability. The basic idea

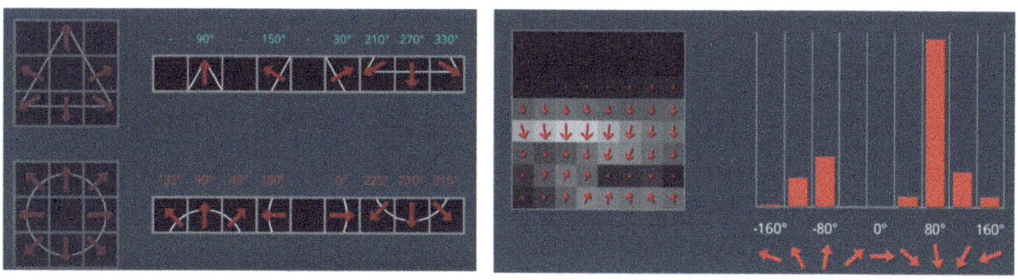


**Fig. 4.9** Gradient calculation of each cell

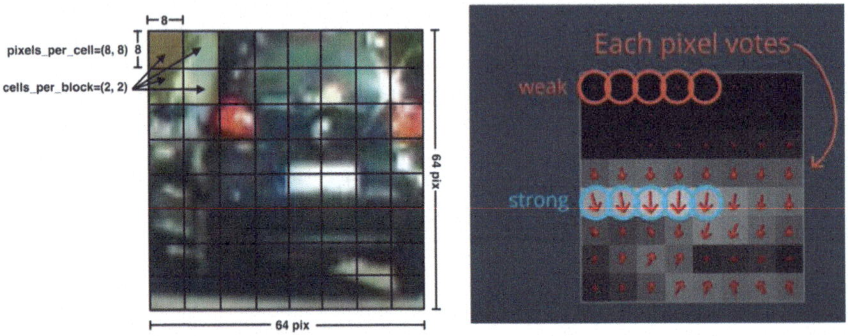


**Fig. 4.10** Schematic diagram of cell units

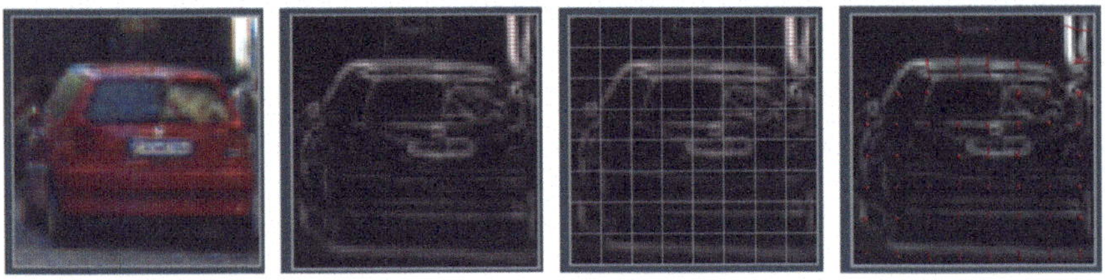

**Fig. 4.11** Illustration of steps taken for HOG feature detection vehicle

can be illustrated in Fig. 4.12. From the diagram, H is the classification line, H1 and H2 are the lines of the sample that are parallel to the classification line, and the distance between them is called the classification interval. The solid point and the hollow point decibel represent two samples. In short, the support vector machine is to produce the middle line, which is a hyperplane in the n-dimensional space.

The optimal hyperplane is calculated as follows:

$$\mathrm{f}(x) = \beta_0 + \beta^T x \tag{4.9}$$

Of which, $\beta$ is the weight vector and $\beta_0$ is the deviation.

By scaling $\beta$ and $\beta_0$, the optimal hyperplane can be represented in infinite different ways. As a convention, among all possible representations of the hyperplane, the following representation is chosen:

$$\left|\beta_0 + \beta^T x\right| = 1 \tag{4.10}$$

Wherein $\chi$ represents the closest training hyperplane. In general, the training example closest to the hyperplane is called a support vector. This representation is called a canonical hyperplane.

Next, using the geometric result, the distance between the point $\chi$ and hyperplane $(\beta, \beta_0)$ is given as:

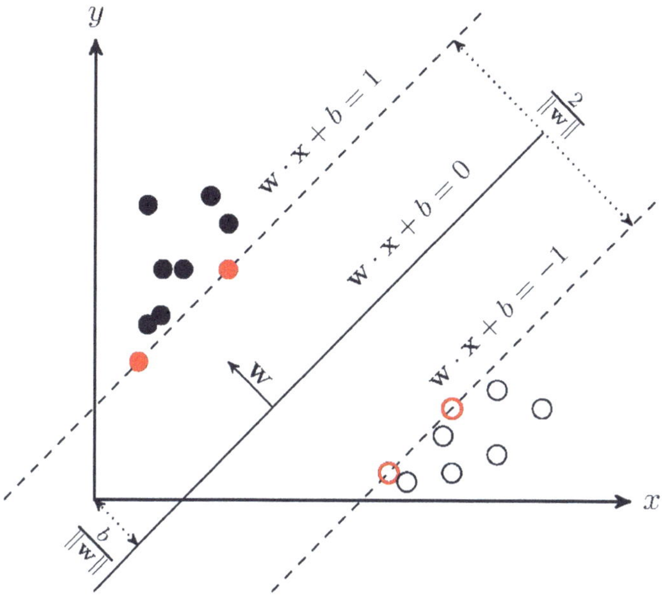


**Fig. 4.12** Schematic diagram of optimal classification planes

$$\text{distance} = \frac{|\beta_0 + \beta^T x|}{||\beta||} \tag{4.11}$$

In particular, for a canonical hyperplane, the numerator is equivalent to 1 and the distance to the support vector is given as:

$$\text{distance}_{support\ vectors} = \frac{|\beta_0 + \beta^T x|}{||\beta||} = \frac{1}{||\beta||} \tag{4.12}$$

$$\text{M} = \frac{2}{||\beta||} \tag{4.13}$$

M is twice the distance from the nearest example.

Finally, the M maximum problem is equivalent to the minimum problem of the L(β) function that is subjected to certain constraints. The requirements of the constrained simulated hyperplane is to correctly classify all the training samples $\chi_i$, and the formula is given as:

$$\underset{\beta, \beta_0}{min}\, L(\beta) = \frac{1}{2}||\beta||^2 subject\, to\, y_i(\beta_0 + \beta^T x_i) \geq 1 \forall i \tag{4.14}$$

#### 4.4.1.2 Image Classification

To find out whether an image includes a certain object, image characterization is the main research content of the object classification. In general, the object classification algorithm globally describes the entire image by manual feature or feature learning method, and then uses the classifier to determine the presence of the specified type of object. The classic image classification algorithm is the proximity algorithm.

(1) K-Nearest Neighbor Algorithm

K-Nearest Neighbor (KNN) classification algorithm is the simplest machine learning algorithm. It uses a method of measuring the distance between different eigenvalues for classification. Its idea is simple: calculate the distance between a point A and all the remaining points and determine the k number of points that are closest to point A. Then find the largest classification of points within the cluster to assign A.

The classification method is based on the nearest one or multiple sample types to determine the classification type. The KNN method is only relevant to a very small number of adjacent samples in class decision making. As the KNN method relies mainly on the surrounding finite number of neighboring samples and is not reliant on the discriminant domain method to determine the category, hence, it is more suitable to use KNN method for cross-over or overlapping sample sets.

In the process of using the model, the appropriate selection of k in the KNN classifier and the use of L1 norm or L2 norm for the distance calculation would require the user's manual selection. These parameters are called the hyperparameters. The training set is divided into a training set and a verification set according to a ratio of 50–90%. The training set trains the model while the verification set adjusts the hyperparameters. For instance, the k hyperparameter is toggled among several different k values in the KNN to train using the training set. Subsequently, the verification set is used to verify for the best k value by plotting the data for analysis before using this k value on the test set to evaluate the algorithm (Figs. 4.13 and 4.14).

#### 4.4.1.3 Image Segmentation

During the image processing process, it is sometimes necessary to segment the image to extract valuable parts for subsequent processing, such as screening feature points, or dividing a part of one or more pictures containing a specific target.

Image segmentation refers to the process of subdividing a digital image into multiple image sub-regions (a collection of pixels, also referred to as super pixels). The purpose of image segmentation is to simplify or change the representation of the image, making the image easier to understand and analyze. More precisely, image segmentation is a process of tagging each pixel in an image that allows pixels with the same tag to have some common visual characteristics.

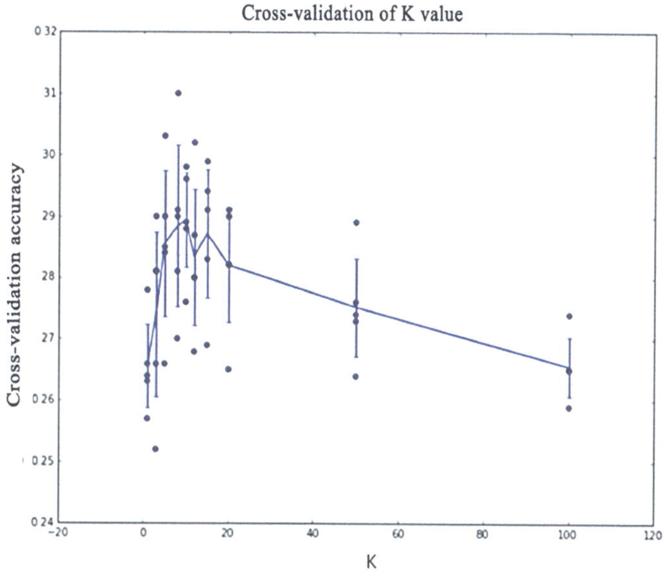


**Fig. 4.13** Results of corresponding KNN classifier using different K value

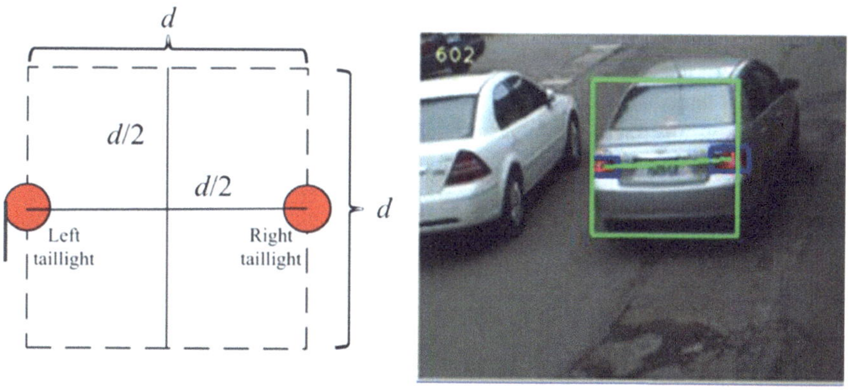


**Fig. 4.14** Application of KNN in vehicle detection

Traditional methods of image segmentation include: (1) Edge detection; (2) Threshold detection.

1. Edge Detection

One of the important ways of image segmentation is via edge detection which detects the abrupt changes in gray levels or structure, and hence, indicating the end of one region and the beginning of another region. This discontinuity is called the edge. Different images possess different gray levels. Generally, there are obvious edges at the boundary. This feature can be used to segment images. The gray value of the pixel at the edge in the image is discontinuous, and this discontinuity can be detected by the derivative. For a stepped edge, its position corresponds to the extreme point of the first derivative, also corresponding to the zero crossing of the second derivative (zero crossing point), so the differential operator is commonly used for edge detection.

Commonly used first-order differential operators include Roberts operator, Prewitt operator and Sobel operator, while second-order differential operators include the Laplace operator and Kirsh operator. In practice, various differential operators are often represented by small area templates, and differential operations are implemented using templates and image convolution. These operators are sensitive to noise and are only suitable for images with less noise and complexity. Since the edge and the noise are both gray-scale discontinuous points, under high-frequency domains, it is difficult to overcome the influence of noise by the direct use of the differential operation. Therefore, the image is smoothed before the edge is detected by the differential operator.

The Roberts operator facilitates the segmentation of low-noise images with steep edges. The Laplacian operator is isotropic. The Roberts operator and the laplacian operator greatly enhance the noise during the implementation process, and hence, worsen the signal-to-noise ratio. Prewitt operators, Sobel operators, etc., are advantageous for the segmentation of images with more noise and of grayscale gradient. The Log operator and the Canny operator are second-order and first-order differential operators with smoothing functions respectively, and the edge detection effect is better. The Log operator is the second derivative of the Gaussian function using the Laplacian operator. The Canny operator is the first derivative of the Gaussian function. It achieves a good balance between noise suppression and edge detection. The Marr algorithm has a smoothing effect on images with more noise, and its edge detection effect is better than the above several operators, but the Marr algorithm causes the image contrast to decrease while smoothing. Kirch algorithm uses an appropriate threshold to binarize a gradient image such that the target and background pixels are below the threshold while most of the edge points are higher than this threshold. At the same time, in order to improve the performance, the watershed algorithm may be introduced for accurate segmentation.

The Hough transform method uses the global characteristics of the image to directly detect the target contour and connects the edge pixels to form a common method for the closed boundary of the region. Under the condition that the shape of the region is known in advance, the boundary curve can be conveniently obtained by the Hough transform to connect the discontinuous boundary pixel points. Its main advantage is that it is less affected by noise and curve discontinuities.

For images with complex gray scale changes and rich details, it is difficult for the edge detection operator to completely detect the edge, and the direct processing effect of the above operator is less ideal when there is noise interference. There are not many examples of this method for segmenting microscopic images because many of the textures or particles in the microscopic image mask the true edges. Although they can be improved by algorithms, the effect is not very good.

Quasi operator (matching model parameters algorithm) principle refers to the fitting of the local gray value in the image with the edge of the model parameters. The advantage of this operator is that it smoothens the noise while detecting the edges which provides better effect for images with higher noise and high textured cellular image. Its disadvantage involves a greater computational overhead, algorithm complexity and higher requirements for edge types which is due to the larger edge structure information recorded in the parameter model (Fig. 4.15).

2. Threshold Detection

The key to the threshold segmentation algorithm is to determine the threshold and to accurately segment the image if a suitable threshold can be determined. If the chosen threshold is too high, too many target areas will be divided into backgrounds, whereas if the threshold is chosen to be too low, too much background will be divided into target areas. After the threshold is determined, the threshold value is compared with the gray value of the pixel. The pixel division can be performed in parallel for each pixel. The result of the division directly gives the image area.

The threshold segmentation must satisfy the hypothesis that the histogram of the image has a distinct bimodal or multi-peak and selects a closed value at the bottom of the valley. Therefore, this method is very effective in segmenting images with large target and background contrast, and it is always possible to define non-overlapping regions with closed, connected boundaries.

The threshold segmentation method is mainly divided into global and local. The closed-value segmentation methods applied at present include the minimum error method, the maximum correlation method, the maximum method, the moment retention method, and the Otsu maximum inter-class variance method. The most widely used is the Otsu's maximum inter-class variance method.

A variety of threshold processing techniques have been developed, including global thresholds, adaptive thresholds, optimal thresholds, etc. The global threshold means that the entire image is segmented using the same threshold and is suitable for images with a clear contrast between the background and the foreground. It is

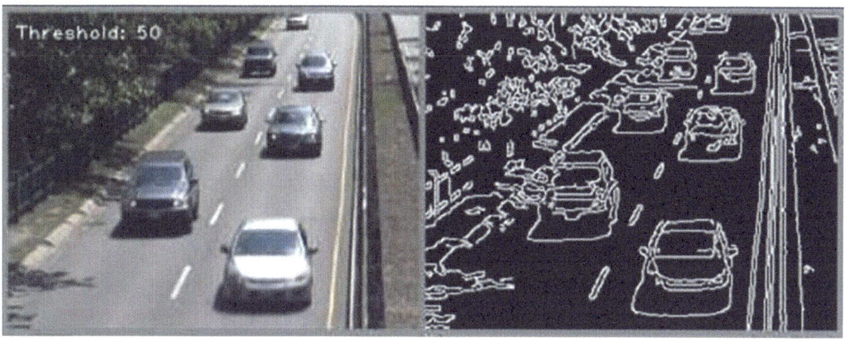


**Fig. 4.15** Schematic view of edge detection by an unmanned car

determined from the entire image: T = T(f). However, this method only considers the gray value of the pixel itself, and generally does not consider the spatial feature, and is therefore sensitive to noise. Commonly used global threshold selection methods include the peak-valley method that uses the image gray histogram, the minimum error method, the maximum inter-class variance method, the maximum entropy automatic threshold method, etc. In many cases, the contrast of the object and the background is not the same everywhere in the image. It is difficult to separate the object from the background with a uniform threshold. At this time, different thresholds may be used for segmentation according to local features of the image. In actual processing, it is necessary to divide the image into several sub-regions according to specific problems to select thresholds, or dynamically select thresholds at each point according to a certain neighborhood range, and perform image segmentation. The threshold at this time is an adaptive threshold. The choice of threshold needs to be determined according to the specific problem, which is generally determined by experiments. For a given image, the optimal threshold can be determined by analyzing the histogram. For example, when the histogram clearly shows a bimodal condition, the midpoint of the two peaks can be selected as the optimal threshold.

The advantages of threshold segmentation include its simple computation, high operational efficiency, high speed, ease of implementing the algorithm and operational efficiency in applications (such as for hardware implementation), and hence is widely adopted. This method is effective for images with large contrast between the target and the background. Additionally, it is always possible to define non-overlapping regions with closed and connected boundaries. However, it is not suitable for multi-channel images and images with little correlation of eigenvalues. It is difficult to obtain accurate results for images where there are no significant gray value differences or a large overlap of gray values of each object. In addition, since the threshold determination mainly depends on the gray histogram and the spatial positional relationship of the pixels in the image is rarely considered, therefore when the background is complicated especially when several research targets overlap on the same background or when there are too much noise signals in the image or the gray value of the target is not much different from the background, it is easy to lose part of the boundary information. Hence, the result obtained by segmentation according to the fixed threshold is inaccurate, resulting in incomplete segmentation and further precise positioning (Fig. 4.16).

## 4.4.2 Machine Vision Based on Deep Learning

Target detection by traditional image processing methods requires designing features and selecting appropriate classifiers. However, this method has a large amount of computation and limited types of recognition targets. For example, SVM classifiers can only classify positive and negative samples. To identify both vehicles and pedestrians, two classifiers must be used, which increases the amount of

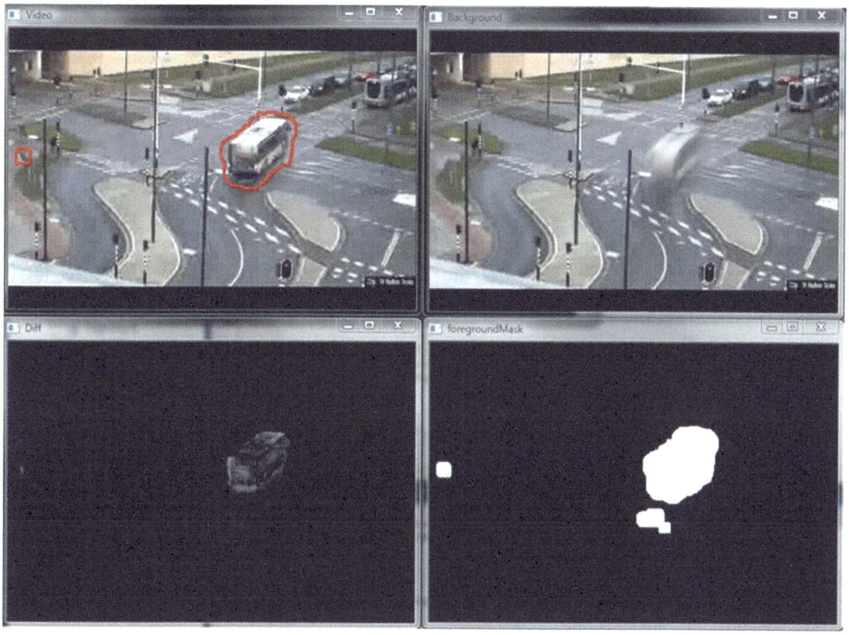


**Fig. 4.16** Threshold segmentation being applied in traffic scenarios

computational complexity and makes it difficult to meet the real-time nature of the algorithm. On the other hand, manually designed traditional image features face stagnation after the HOG and DPM features which has not been able to achieve breakthroughs in accuracy until the convolutional neural network is proposed.

In 2012, scholars applied the deep convolutional neural network for the first time in the field of target recognition. This breakthrough practice improved the classification accuracy record in ImageNet dataset by about 10%. They further deepen and widen the convolutional neural network to achieve more complex target recognition. In addition, the ReLU activation function and maximum pooling method are applied in a pioneering manner, and the Dropout training method and LRN layer are designed to make parameter transmission more rapid and effective. The application of these methods enables the training of the deeper and wider network of the team, which proves the potential and advantages of deep learning.

In the neural network, the structure of each unit is as shown in Fig. 4.17, and its corresponding formula is:

$$h_{W,B}(x) = f\left(W^T x\right) = f(\sum_{i=1}^{n} W_i x_i + b) \tag{4.15}$$

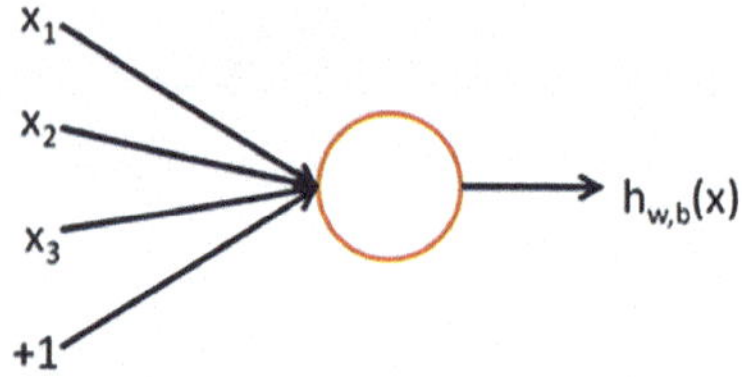


**Fig. 4.17** Neural network model

Wherein, the unit can also be called a logistic regression model. When a plurality of units are combined to have a hierarchical structure, a neural network model is formed.

Figure 4.18 shows a neural network with an implicit layer.

The corresponding formula is as follows:

$$\begin{aligned} a_1^{(2)} &= f(W_{11}^{(1)}x_1 + W_{12}^{(1)}x_2 + W_{13}^{(1)}x_3 + b_1^{(1)}) \\ a_2^{(2)} &= f(W_{21}^{(1)}x_1 + W_{22}^{(1)}x_2 + W_{23}^{(1)}x_3 + b_2^{(1)}) \\ a_1^{(2)} &= f(W_{11}^{(1)}x_1 + W_{12}^{(1)}x_2 + W_{13}^{(1)}x_3 + b_1^{(1)}) \\ h_{W,b}(x) = a_1^{(3)} &= f(W_{11}^{(1)}a_1^{(2)} + W_{12}^{(1)}a_2^{(2)} + W_{13}^{(1)}a_3^{(2)} + b_1^{(2)}) \end{aligned} \tag{4.16}$$

Those that are more similar can be expanded to 2, 3, 4, 5, … hidden layers, the number of neurons in each layer can also be expanded.

The training method of neural network is similar to Logistic, but because of its multi-layered nature, it is also necessary to use the chain-guided rule to deduct the

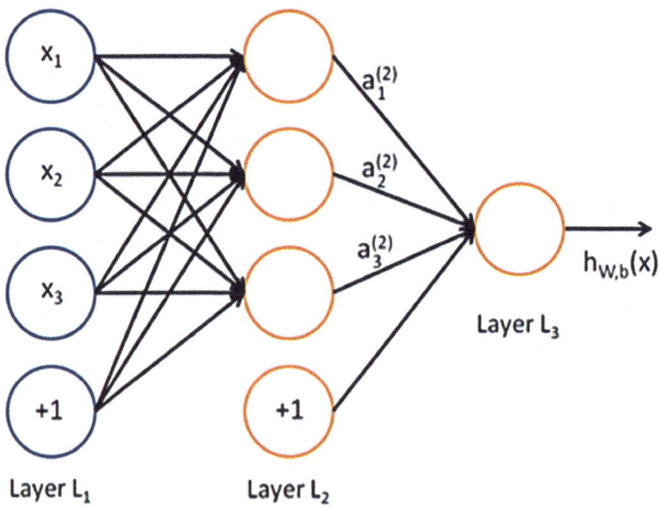


**Fig. 4.18** Neural Network model with an implicit layer

nodes of the hidden layer, that is, the gradient descent and chain derivation rule, also known as the reverse propagation.

Convolutional neural networks differ from the ordinary neural networks in the sense that the convolutional neural network consists of a feature extractor that includes a convolutional layer and a subsampling layer. In the convolutional layer of a convolutional neural network, one neuron is only connected to a portion of the adjacent layer neurons. There are usually several feature maps in the convolutional layer. Each feature plane consists of a number of rectangularly arranged neurons. The neurons of the same feature plane will share the weights. The weights shared here are known as the convolutional kernel. The convolution kernel is generally initialized in the form of a random fractional matrix. It will learn to obtain reasonable weights during the training process of the network. The immediate benefit of shared weights (convolution kernels) is the reduction of connections between layers of the network while reducing the risk of overfitting. Subsampling is also called pooling. It usually has two forms: mean pooling and max pooling. Subsampling can be seen as a special convolution process. Convolution and subsampling greatly simplify the model complexity and reduce the parameters of the model. The basic structure of the convolutional neural network is shown in Fig. 4.19.

The convolutional neural network consists of three parts. The first part is the input layer. The second part consists of a combination of n number of convolutional layers and pooled layers. The third part consists of a fully connected multi-layer perceptron classifier.

In order to reduce the number of parameters of the multi-layer network, the convolutional neural network adopts two methods. The first method is local sensing field. It is generally believed that people's perception of the outside world is from local to global, and the spatial connection of images is also a close connection of local pixels, while the correlation of pixels far away is weak. Therefore, it is not

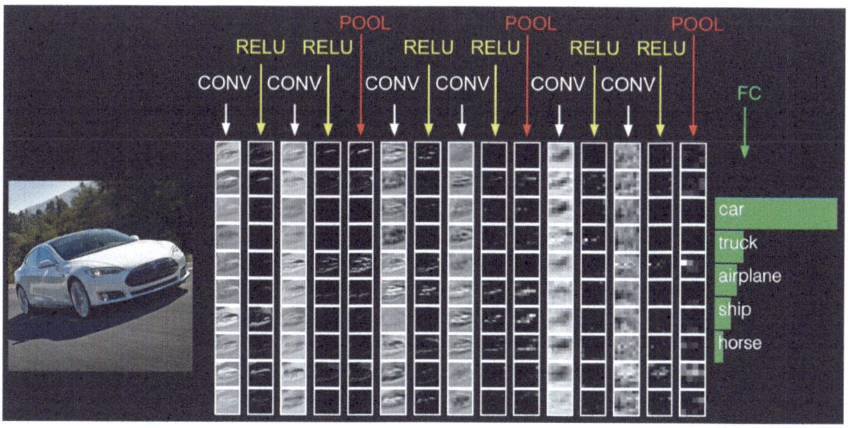


**Fig. 4.19** Basic structure of the convolutional neural network

necessary for each neuron to perceive the global image. It only needs to perceive the local part, and then integrate the local information at a higher level to obtain global information. The idea of part of the network is also inspired by the structure of the visual system in biology. Neurons in the visual cortex locally accept information (i.e., these neurons only respond to stimuli in certain areas). As shown in Fig. 4.20: the left picture illustrates the full connection, the right picture indicates the local connection.

The second method employs weight sharing. Weight sharing refers to the same convolution layer operation where different located neurons will share the same set of parameters. The concept is that if the convolution operation is treated as a mode of extracting features, then this method would be independent of the location. That is, the statistical characteristics of a part of the image are the same as the other parts. This also means that the characteristics of the network learning in this part can also be used in another part, so the same learning characteristics can be applied for any position in the image. Figure 4.21 shows the result of convolving a 5 × 5 image with a 3 × 3 convolution kernel, in which each convolution result is derived from the same set of convolution parameters.

The above image illustrates for only one convolutional process. Obviously, such feature extraction is not sufficient. Multiple convolution kernels can be added to the convolutional neural network, such as 32 convolution kernels so that 32 features can be learned. When there are multiple convolution kernels, the result is as shown below (Fig. 4.22).

Since features that are useful in one image region are most likely to be equally applicable in another region, in order to describe a large image, it is necessary to reduce the output dimension to aggregate statistics of the features at different locations. For instance, the mean or maximum value could be computed for a specific feature in an area of the image. It has a much lower dimension than the extracted features, while also improving the robustness of the results and avoiding overfitting. This operation is called the polymeric pooled (Pooling), sometimes referred to as the average or best pool of the large pooled (depending on the method of calculating the pool) (Fig. 4.23).

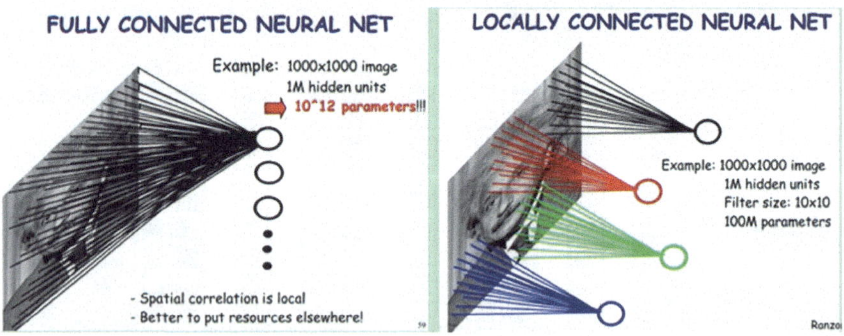


**Fig. 4.20** Neurons in the visual cortex

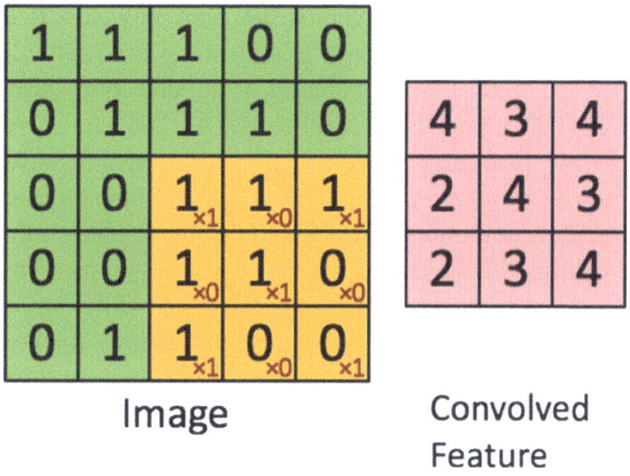


**Fig. 4.21** Results after image convolution

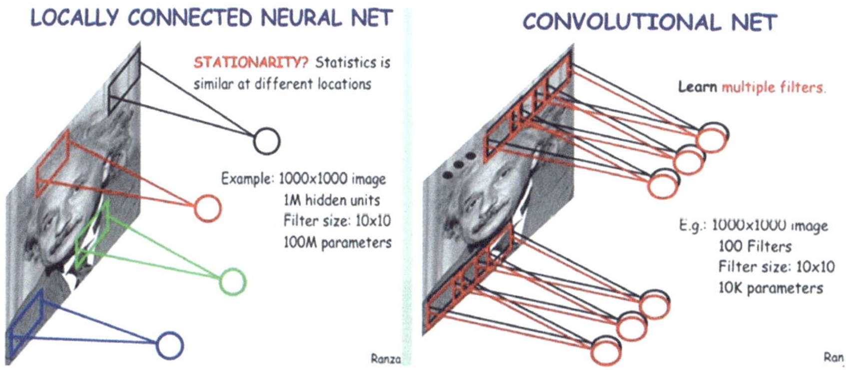


**Fig. 4.22** Multiple convolution kernels

**Fig. 4.23** Pooling operation

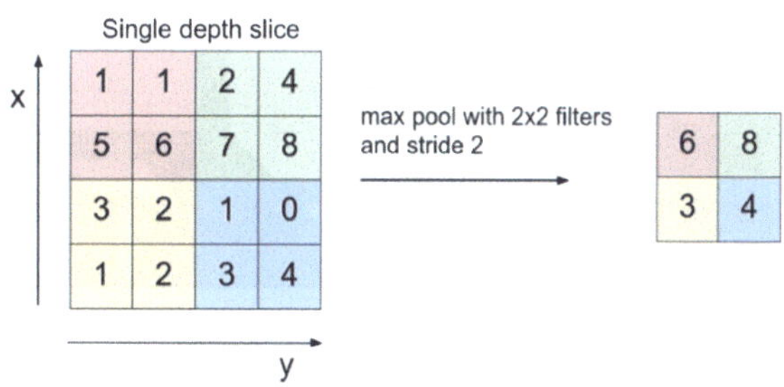


As compared with the traditional image processing methods, the convolutional neural networks can identify more features, have higher accuracy, and can easily realize simultaneous recognition of multiple targets. After being introduced, they have received extensive attention and are developing rapidly.

Since the convolutional neural network (CNN) can only classify the input target, the traditional sliding detection window is still being used when performing the target detection on the entire image, resulting in the detection speed of being very slow. The Region CNN (R-CNN) was proposed in 2013 to solve this problem. They use the Selective Search method to pre-generate possible candidate windows instead of the sliding window method of traditional target detection. Selective Search uses an over-segmentation method to segment the image into small regions. By comparing the similarities of the small regions, the regions with high similarity are combined to obtain a large number of regions with different levels, assuming that these regions are likely to exist. These sets of regions are normalized to the same size and then imported into the CNN network to extract features. The obtained feature vector is input into the SVM classifier for prediction, and a regression network for adjusting the size and position of the frame is also designed. The structure of R-CNN is shown below (Fig. 4.24).

However, there are thousands of regions obtained through Selective Search, which still cannot effectively reduce the computational complexity of the algorithm. In 2015, the SPP layer method provided a solution. The SPP layer is generally used in the front of the fully connected layer. Since the full connection layer requires the same dimension, the SPP layer can normalize the features extracted from different sizes to the same dimension. The introduction of the SPP method avoids the distortion introduced by the region scaling as it eliminates the scaling operation which shortens the detection time. In the same year, some scholars proposed Fast R-CNN

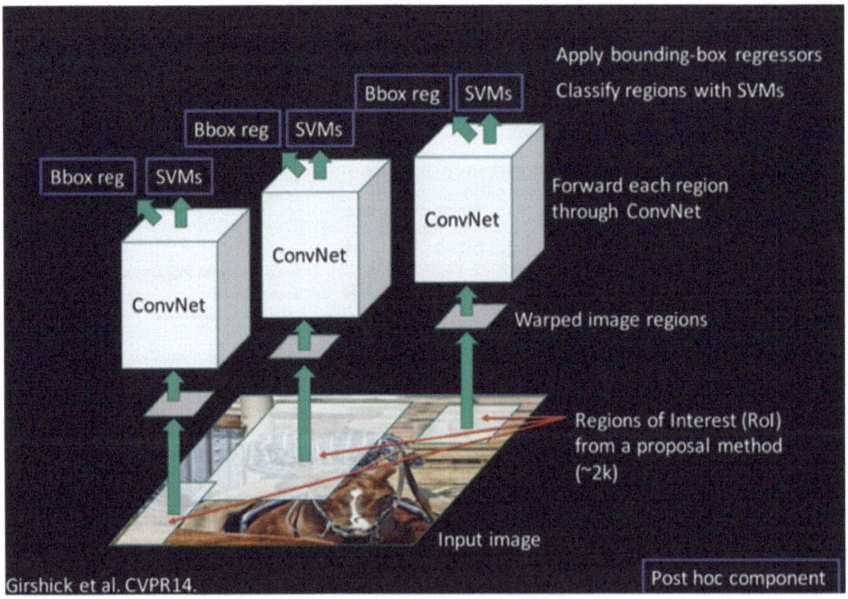


**Fig. 4.24** R-CNN structure

on the SPP method. Fast R-CNN no longer inputs thousands of regions into the CNN extraction feature. Instead, the original image is completely input into the convolution network, and then the volume is replaced. The corresponding feature of the product maps the candidate region to the convolution feature map of the CNN output, and inputs the mapped feature map region into the SPP layer. The prediction result output no longer uses the SVM classifier but combines the SPP layer and the two fully connected layers in which one is used for classification and the other is for adjusting the size and position of the classification box. Thereby, greatly reduce the computation volume, running time and memory required for the operation. Figure 4.25 illustrates the Fast R-CNN structure.

Fast R-CNN omitted the need for thousands of candidate regions to run through the convolutional network, avoiding the most time-consuming traditional image processing step by the Selective Search method. Then the quick appearance of Faster R-CNN further improved the processing. Faster R-CNN proposed the Region Proposal Networks method which performs window sliding via a set of windows of different sizes and aspect ratios on a feature image output by the convolutional layer. These windows are regarded as candidate regions to adjust for classification and detection of frame size and position. The output from the Region Proposal Networks that is regarded as candidate regions is imported into the Fast R-CNN to replace the Selective Search. In Faster R-CNN, Region Proposal Networks shares the feature of convolutional network extraction with the Fast R-CNN. The SPP layer is connected after the Region Proposal Networks. Figurc 4.26 shows the structure of Faster R CNN.

After the appearance of the Faster R-CNN, a faster operation called YOLO is introduced. The main concept of YOLO involves the breaking up of an image into numerous grids. If a target centre is located in one of the grids, this grid will be used to classify the target. The return target's classification and frame detection position

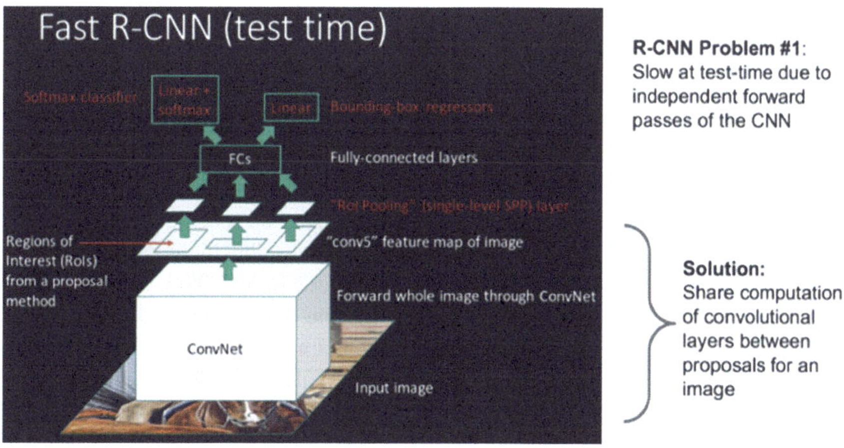


**Fig. 4.25** Fast R-CNN structure

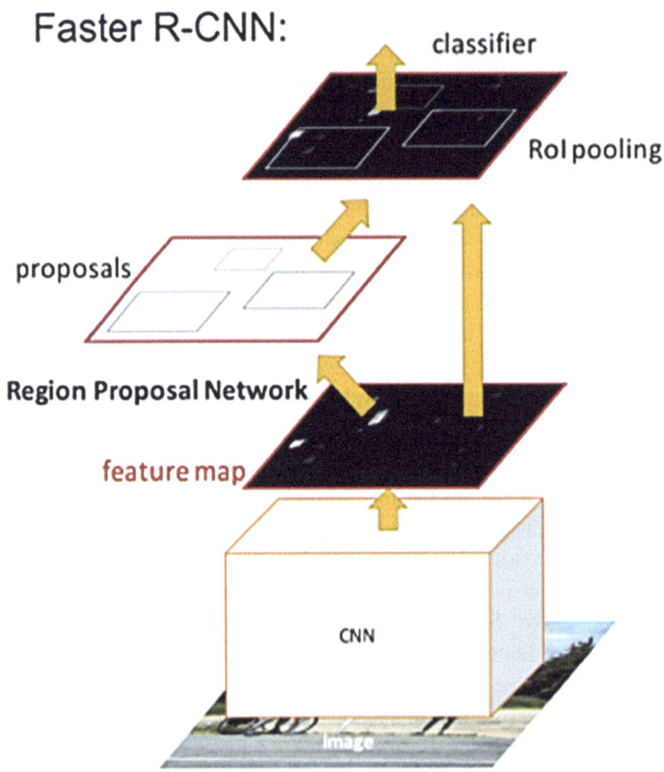


**Fig. 4.26** Faster R-CNN structure

will be adjusted in the output layer. The greatest advantage of YOLO is its ability to fulfil real-time detection based on its operational speed. The SSD, after introducing YOLO, balances between fast detection speed and high accuracy well, achieving a breakthrough in the performance of the target detection algorithm. Figure 4.27 shows the recognition effect of SSD on multiple targets.

Convolution neural network is widely used in the beginning to detect and identify targets. Another mission of computer vision, i.e. semantic segmentation, also began to apply deep learning. The full convolution neural network (FCN) that was introduced in 2014 is the first pixel-level segmentation scheme that uses a deep convolutional neural network method which achieves the highest segmentation accuracy at that time. In order to output the segmented image, the network needs to output a two-dimensional result. FCN abandons the fully connected layers in the convolutional neural network, and replaces all of them with convolutional layers, so that the output of each layer is a two-dimensional feature map. The result is obtained by interpolating the feature map output from the convolutional neural network to the input picture size.

**Fig. 4.27** SSD recognition of multiple targets

After the surface of the FCN, the convolutional neural networks for semantic segmentation continue to evolve. In 2014, Google researchers presented the Deeplab. Deeplab uses the convolution neural network and the conditional random field (CRF) structure, and introduced the hole convolution which can obtain a larger and denser feature map without changing the neurons in every convolutional layer. The conditional random field is a statistical method that uses a probabilistic model to achieve a more subtle segmentation at the target boundary. In order to further improve the Deeplab, the CRF can be trained from end-to-end. Subsequently, studies have also divided the CRF into several steps and then approximated it with the RNN network. In 2016, the Deconv Net was proposed. This method is opposite to the forward transfer process. The deconvolution and anti-pooling layers are introduced to restore the feature map to the original image size and achieve the segmentation effect. In 2016, Google researchers proposed Deeplab V2. Based on V1, the ASPP structure was proposed. The original feature maps were pooled at different scales, and the results of each scale were combined to achieve better segmentation results.

In 2016, the deep residual network is born which greatly increased the effectiveness of the gradient transfer in the deep network, constructs more than hundreds of neural networks and achieve the best results in image recognition task. Semantic segmentation is also introduced into this structure. In 2016, the authors of RefineNet designed a chain-like residual network architecture which is made up of multi-scale integrated complex network structure. Similarly, in 2016, PSPNet was proposed which uses a residual network with hole convolution. It also uses convolution processes of different dimensions, then combines the results of each scale and adds a loss to the training to make the training more effective. In 2017, large convolutional kernels were proposed to analyze semantic segmentation problems from a different perspective. The authors believe that the image segmentation constitutes two tasks, namely classification and positioning. The results of the classification task require the position of an object to remain invariant. This is in contradiction to the positioning task. Hence, to resolve the contradiction, the receptive field area is expanded to increase the positioning sensitivity. On this basis, the scheme of large convolution kernel is proposed. In order to reduce the high volume of calculation caused by the large convolution kernel, the $N \times 1$ and $1 \times N$ convolutions are used in an attempt to approximate to the $N \times N$ convolution effect. In 2017 Deeplab V3 is introduced which further optimizes the ASPP structure in V2. Additionally, it introduced a series of multi-dimensional hollow convolutions. At the same time, it uses the Batch Normalization method to further improve the accuracy of segmentation which is the best performing image semantic segmentation method at that time. The images below depict the image semantic segmentation effect by Deeplab V3 (Fig. 4.28).

**Fig. 4.28** Deeplab V3 image semantic segmentation effect

## 4.5 Machine Vision Application

### *4.5.1 Drone Application*

Computer vision, which was previously limited to computer science research and machine vision technology for specific industry applications, has also moved to the forefront in the AI era, providing the 'eyes' for countless smart devices. Machine vision technology provides great support in the background whether for drone delivery or automated flying. In the era of artificial intelligence, the three most popular words appearing are drones, autonomous vehicles and robots. What are the commonalities of these unmanned smart devices? Firstly, these unmanned devices need to have a "brain", which in this case uses computer instead of the human brain to process large amounts of complex information. Secondly, these unmanned devices need the "eyes" to sense the environment and respond in a timely and correct manner. The "brains" of these smart machines are made up of a set of high-performance CPU chips whose "eyes" are realized by cameras, visual processors (VPUs) and proprietary software processing systems. The motivation behind this "eye" is what we call computer vision or machine vision technology.

Machine vision is similar to computer vision in its basic principle. It can be said to be a branch of system engineering for the computer vision in automated inspection and industrial control. Unlike the computer vision, machine vision focuses on integrating existing technologies in new ways and is used to solve real-world image processing problems. A basic machine vision system consists primarily of a light source, a camera, a vision processor, and an output component. Its main application is in automated inspection and industrial robot guidance systems. The drones developed in recent years are also an emerging application market for machine vision.

The first company with the approval to possess the UAV autopilot had launched a model of industrial UAV which realizes unmanned flight and is equipped with military grade avionics for precise control and conducts pre-programming to carry out tasks. It can automatically start and land, and perform half an hour of flying without human control. The UAV is able to deliver real-time video images via its cameras. Hence, it can be used for air surveillance and patrols to provide security services. It can replace humans for missions in the event of an emergency, and hence, avoid dangers for the human operator (Fig. 4.29).

Civil drones are widely used. In the past, it is widely treated as a toy or a consumer machine. However, now they are used for aerial photography, videography and geological surveys which include security surveillance, traffic monitoring, law enforcement and plant protection such as agriculture and emergency relief. One of the drone series uses the precision hovering technology of binocular stereo vision, which breaks through the limitations of optical flow positioning and even achieves accurate hovering without GPS signals when flying outdoors at high altitudes. It can sense the environment at 30 m ahead of itself in real time, and can automatically hover or bypass at 15 m before obstacles. Thus, greatly improve the

**Fig. 4.29** Industrial drone

safety of the flight. In addition, ground image can be captured by the camera. It can accurately return to the take-off location when returning. This precision hover function is realized thanks to its machine vision processing chip and the accompanying machine vision intelligent algorithm system (Fig. 4.30).

In the AI era, computer vision and machine vision technology, which was previously limited to computer science research and industry-specific applications, began to progress, providing the 'eyes' for countless smart devices. With the rapid development of the emerging smart devices such as the drones and driverless machinery, the distinction between computer vision and machine vision has become blurred. We can see more intelligent 'eyes' scanning and collecting information of the surrounding environment whether in a factory or store, or on the highway or in the air. The popularity of these smart devices not only liberate people of manual labor but also 'eye-power'.

**Fig. 4.30** Civil drone

## 4.5.2 Unmanned Vehicles Application

In the field of automated driving, visual sensor plays an important role in sensing the environment. In the driving assistance systems that had been applied, many systems such as the LKA, LDW, FCW, etc. had made use of visual sensors such as the cameras.

As an important part of driverless driving, the Lane Keeping Assist System (LKA) is often placed in the pre-development phase. First with driver assistance through LKA, and then gradually move towards driverless through gradual improvement. LKA can be broadly divided into two categories: lane keeping (LK) and lane centering (LC). The difference lies in the control system, control target and the degree of control system intervention. The control target of the lane departure warning system is to keep the vehicle in lane with less assistance. The control target of the lane centering assist system is to keep the vehicle near the centerline of the lane, which provides more assistance in comparison to the lane departure warning system (Fig. 4.31).

The camera can obtain the distance between the detected lane line and the vehicle body through the imaging principle and the corresponding calibration operation. On the other hand, the angle between the lane line and the traveling direction of the vehicle can also be obtained, thereby estimating the time of deviation from the lane. Based on the different control strategies, the distance or estimated time becomes the basis of the LKA control system. The accuracy in detecting the lane line and calculation of the distance or estimated time is the premise and difficulty of the LKA technology. The uniqueness of the lane line is often detected via the camera. Hence, the camera-based lane line detection forms the mainstream scheme of the LKA technology.

The lane departure warning system (LDW) similar to the lane departure assist system (LDA) is also mainly based on the camera-based lane line detection

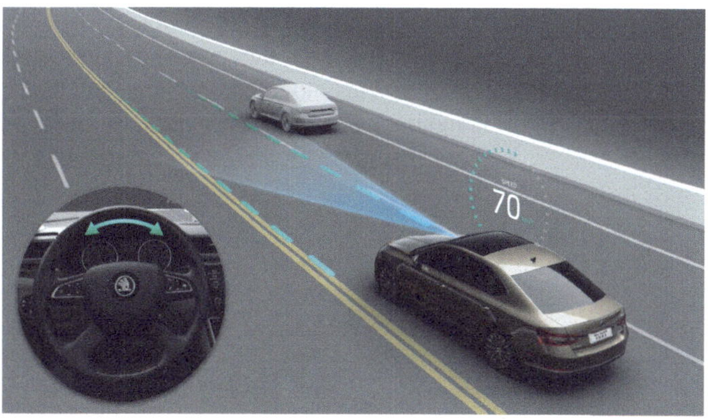


**Fig. 4.31** Lane keeping assistance system

technology. Unlike the LKA active intervention vehicle control, the LDW mainly provides lane departure warning information.

Another application of visual perception is the Forward Collision Warning System (FCW). In the FCW system, the camera is responsible for detecting and recognizing the forward obstacles. Based on the imaging principle of the camera, the distance between the obstacle and the vehicle body is being judged. The estimated collision time is calculated according to the distance and the speed of the vehicle. The estimated collision time is used as the basis for judging whether or not to issue an early warning. In the forward collision warning system, the millimeter wave radar is also a commonly used sensor. Since the millimeter wave radar can obtain the target speed information more accurately, and the camera can judge the target category, the fusion of information from both devices forms the current mainstream solution for the forward collision early warning system.

Similar to the forward collision warning system, the Automatic Emergency Brake System (AEB) also requires the detection of forward obstacles. The AEB system detects and recognizes the front vehicle by a camera or radar. When there is a possibility of a collision, the audio alert and the warning light are used to remind the driver to perform the braking operation to avoid collision. If there is no braking action performed by the driver and the system judges that it is impossible to avoid a rear-end collision, and an automatic braking measure is taken to reduce or avoid the collision. At the same time, the AEB system also includes dynamic brake support, which is reinforced when the driver's force on the brake pedal is insufficient to avoid an imminent collision. Figure 4.32 shows the effect of the AEB system.

### 4.5.3 Unmanned Boats Application

Pertaining to the environment sensing technology in unmanned boats, it mainly includes visual perception technology, infrared sensing technology, radar sensing technology and underwater acoustic sensing technology. Compared with the other sensing technologies, optical vision-based sensing technology is able to more effectively distinguish surface targets because the optical images contain more comprehensive target area details. Cameras mounted on unmanned boats are often used as an important aid for navigation. The digital camera has unique advantages

**Fig. 4.32** Performance of the AEB system

as an auxiliary sensing device. The captured video material has high resolution, low cost and easy implementation. Research on camera-based surface target detection and tracking technology is beneficial in the effective extraction of information of large-area coastal areas, sailing vessels, low-lying suspended obstacles (such as bridges near the coast and branches) and surface floating obstacles (such as floating objects and floating debris) during the operation of surface unmanned boats. It also helps to autonomously avoid the obstacles, thereby avoid shipwrecks or ships collision and improve the survival and operational capabilities.

At the same time, it is also one of the supporting technologies for accomplishing anti-mine warfare, maritime safety, anti-submarine warfare and surface combat missions. It is one of the key technologies to ensure that unmanned boats on the surface do not need to return to the mother ship or rely on personnel to remotely perform multiple tasks.

The camera is mainly used in the unmanned boat for the identification of the surrounding environment to achieve obstacle avoidance. The installation position of the camera on the unmanned boat can be divided into two categories:

(1) Around the hull of the unmanned boat: Used to collect panoramic images, which can detect obstacles around the hull;
(2) Above the ship: The surrounding bird's-eye view image is taken to detect long-distance targets and the detection distance for other ships, coasts and other large targets which can reach several kilometers.

Figure 4.33 shows a diagram of the camera that is installed on an unmanned boat. Since the image captured by the camera is highly susceptible to changes in the illumination, the sun's glare reflected from the water surface sometimes interferes

**Fig. 4.33** Diagram of a camera installed on an unmanned boat (red colored depicts the camera at the hull; black colored depicts bird's eye camera)

**Fig. 4.34** U.S. Navy's "Seafox" series of unmanned boats

with the camera, which also places high demands on the image processing algorithm.

Figure 4.34 shows the "Seafox" series of unmanned boats developed by the US Navy. This model of unmanned boat is equipped with environmentally sensitive sensors such as the digital zoom infrared camera, the digital zoom white color camera, and the $3 \times 70°$ navigation camera. These devices allow this model of USV to possess the ability of remote monitoring and significantly widen the unmanned boat's operating range which includes the inland operation, sea interception operation, marine environment awareness, port security, automatic operation, etc.

## Bibliography

1. Chen H (2009) Classification of universal cameras. Telev Technol 33(5):127–128
2. Gao X, Zhang T (2017) Visual SLAM XIV: from theory to practice. Electronics Industry Press, p 3
3. Gao Z, Zhang Y, Sun P et al (2017) Review of research on unmanned ships. J Dalian Marit Univ 43(2):1–7
4. Zeng W (2013) Research on water surface target detection and tracking of unmanned boats based on optical vision. Harbin Engineering University
5. Zhang Y (2015) Research on visual target image recognition technology for surface unmanned boats. Harbin Engineering University
6. Winner H, Hakuli S, Lotz F et al (2015) Handbook of driver assistance systems
7. Bay H, Ess A, Tuytelaars T et al (2008) Speeded-up robust features. Comput Vis Image Underst 110(3):404–417

8. Jia Y (2000) Machine vision. Science Press, p 4
9. Zhang J (2006) Research on image acquisition and processing system based on machine vision. Chengdu University of Technology
10. Zhang Y, Wu X, Ma T et al (2006) Research on image acquisition and recognition of parts based on machine vision. Inf Res 32(4):29–31
11. Ma J (2017) Vision-based unmanned aerial vehicle target recognition and tracking control. Dalian Maritime University
12. Wang D (2016) Research on vision-based UAV detection and tracking system. Harbin Institute of Technology
13. Li X (2017) Research on target tracking and localization algorithm of drone based on computer vision. Yanshan University
14. Ding M (2006) Research on autonomous landing method of drone based on computer vision. Nanjing University of Aeronautics and Astronautics
15. Yang F (2008) Vision-based research and implementation of UAV autonomous landing system. Tsinghua University
16. Huang Z (2018) Autonomous landing system of UAV based on machine vision. China Strategic Emerging Industries, p 4
17. Zou Y (2016) Research on autonomous landing technology of drone based on machine vision. Jiangsu University
18. He J, Li B (2018) Research on autonomous landing system of UAV based on machine vision. Science and Technology Innovation, p 11
19. Shilin婕 (2015) Research on UAV visual aid landing algorithm. Xidian University
20. Li G (2016) Research on UAV power line inspection technology based on machine vision. Anhui University of Science and Technology
21. 肖鹏 (2017) A machine vision based unmanned aerial vehicle obstacle avoidance system, CN205910595U[P]

# Chapter 5
# Infrared Sensors and Ultrasonic Sensors

## 5.1 Introduction

In addition to the environmentally-aware sensors that were introduced in the previous chapters, there are some sensors that are widely used in unmanned systems and everyday life. Among them, infrared detection sensors and ultrasonic sensors are also important sensing devices in unmanned systems.

Infrared (IR) detection is a technique that uses an instrument to receive infrared rays emitted or reflected by a detected object to grasp the position of the measured object. Infrared sensor has good environmental adaptability, good concealment, strong anti-interference ability, can identify camouflage target to a certain extent, and has the characteristics of small size, light weight and low power consumption, which can make up for the shortcomings of other sensors. Hence it is highly important in an unmanned system. However, due to the high cost, it has not been adopted widely.

The ultrasonic sensor is a sensor that converts an ultrasonic signal into other energy signal (usually an electric signal). Ultrasonic waves are mechanical waves with a vibration frequency higher than 20 kHz. It has the characteristics of high frequency, short wavelength, small diffraction phenomenon, especially good directionality, and can be directional and propagated. Ultrasonic has excellent penetration in liquids and solids, especially in opaque solids. When an ultrasonic wave hits an impurity or an interface, it will undergo significant reflection to form a reflected wave, which can produce a Doppler effect when it hits a moving object. Ultrasonic sensors are widely used in unmanned systems.



X. Bi, *Environmental Perception Technology for Unmanned Systems*, Unmanned System Technologies, https://doi.org/10.1007/978-981-15-8093-2_5

## 5.2 Infrared

The infrared is part of the solar spectrum. Its greatest characteristic is its photothermal effect and its ability to radiate heat. It is the largest spectral region of photothermal effect. The infrared is invisible and like all the other electromagnetic waves, it possesses the ability to reflect, refract, scatter, interfere and absorb. Infrared propagation velocity in vacuum is $3 \times 10^8$ m/s. Infrared in the medium will produce propagation attenuation which is the largest in metal. However, infrared is able to penetrate most of the semiconductors and plastics. Most of the liquids are able to highly absorb the infrared. The degree of absorption will vary in different types of gas. The atmosphere of the different infrared wavelengths have different absorption bands. Analysis shows that, for a wavelength between 1 and 5 μm, 8 and 14 μm infrared region has a comparatively larger transparency and is able to better penetrate the atmosphere. Any object in the natural environment with a temperature above absolute zero can produce infrared radiation. The infrared photothermal effect and energy intensity are different on different objects. For instance, black body (able to fully absorb the infrared radiation that is projected onto its surface), mirror (able to fully reflect the infrared radiation from an object), transparent body (able to fully penetrate the infrared radiation object) and gray body (able to partially reflect or absorb the infrared radiation) will produce different photothermal effects (Fig. 5.1).

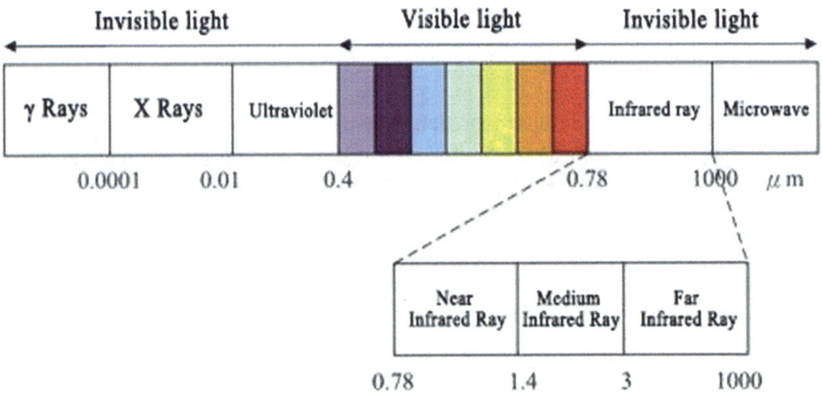


**Fig. 5.1** Schematic diagram of infrared spectrum

## 5.3 Classification of Infrared Sensor

The infrared sensor is a sensor that uses infrared rays as a medium for data processing. According to the different ways of transmission, infrared sensors can be divided into active and passive.

1. **Active Infrared Sensor**

The transmitter of the active infrared sensor emits a modulated infrared beam that is received by the infrared receiver to form a warning line of infrared beams. When it encounters leaves, rain, small animals, snow, sand, and fog, it will not be alarmed. If a person or an object of a considerable volume causes a blockage, an alarm will occur.

Active infrared detector technology mainly adopts a single transmit and receive strategy which is a method of linear prevention. Currently it has developed from the original single beam to multiple beams and is able to perform double transmit and receive function which could minimize the false alarm rate and hence enhance the stability and reliability of the product. As the infrared is an excellent non-coherent environmental factor detection medium (for sound, lightning, vibration, various types of artificial light sources and electromagnetic interference sources in the environment), and at the same time it is also a product of target factors with good coherence, and hence the active infrared sensor is further promoted for use and application.

2. **Passive Infrared Sensor**

Passive infrared sensors operate by detecting the infrared rays that are emitted by the human body. The sensor collects the external infrared radiation onto the infrared sensor. Infrared sensors typically use pyroelectric elements that discharge a current outward when they receive a change in the temperature of the infrared radiation. This sensor is aimed at detecting human radiation. Therefore, the radiation-sensitive elements need to be very sensitive to infrared radiation wavelength of about 10 μm. In order to be sensitive to the infrared radiation of the human body, its radiation surface is usually covered with a special filter, so that the environmental interference is significantly controlled.

A passive infrared sensor consists of two pyroelectric elements connected in series or in parallel. Moreover, the two polarization directions are exactly opposite. The environmental background radiation has almost the same effect on the two pyroelectric elements, so that the discharge effect is mutually canceled, so that the detector has no signal output. Once the intruder enters the detection area, the infrared radiation of the human body is focused by the partial mirror and received by the pyroelectric element, but the two pieces of pyroelectric elements receive different heat and the pyroelectricity is different, and hence cannot be mutually cancelled and a signal is transmitted after processing.

According to different energy conversion methods, infrared sensors can be divided into photon and pyroelectric.

1. **Photon Infrared Sensor**

Photonic infrared sensors are sensors that operate using the photon effect of infrared radiation. The so-called photon effect means that when infrared rays are incident on some semiconductor materials, the photon flux in the infrared radiation interacts with the electrons in the semiconductor material, changing the energy state of the electrons, thereby causing various electrical phenomena.

By measuring the change in the electronic properties of the semiconductor material, the intensity of the corresponding infrared radiation can be known. The photon detector types mainly include internal photodetectors, external photodetectors, free carrier detectors, and QWIP quantum well detectors (Fig. 5.2).

The main characteristics of the photon detectors include high sensitivity, fast response and high response frequency, but the disadvantage is that the detection band is narrow and generally working at low temperature (to maintain high sensitivity, often liquid nitrogen or thermoelectric cooling, etc. is used to cool the photon detector to a lower operating temperature).

2. **Pyroelectric Infrared Sensor**

The pyroelectric infrared sensor uses the thermal effect of the infrared radiation to cause the temperature change of the component itself to detect certain parameters, and the detection rate and response speed are not as good as the photon sensor.

However, since it can be used at room temperature, the sensitivity is independent of the wavelength, and so the application field is wide. The pyroelectric type infrared sensor utilizing the pyroelectric effect of ferroelectric is highly sensitive and has been widely used (Fig. 5.3).

The pyroelectric effect of some insulating materials when heated result in an equal number of oppositely charged charges to be produced across the crystal as the temperature rises. This phenomenon of polarization due to thermal changes is called pyroelectric effect. The pyroelectric effect has been used in pyroelectric infrared sensors for nearly a decade. A crystal that produces a pyroelectric effect is called a pyroelectric material, which is also called a pyroelectric element. Common materials for thermoelectric elements include single crystal, piezoelectric ceramics, and polymer films.

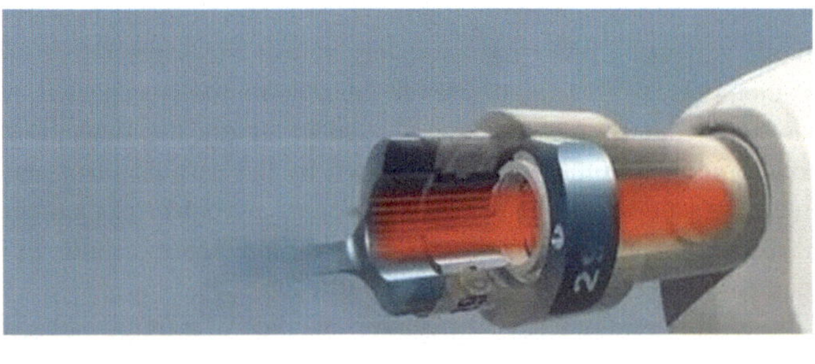

**Fig. 5.2** Photon infrared sensor

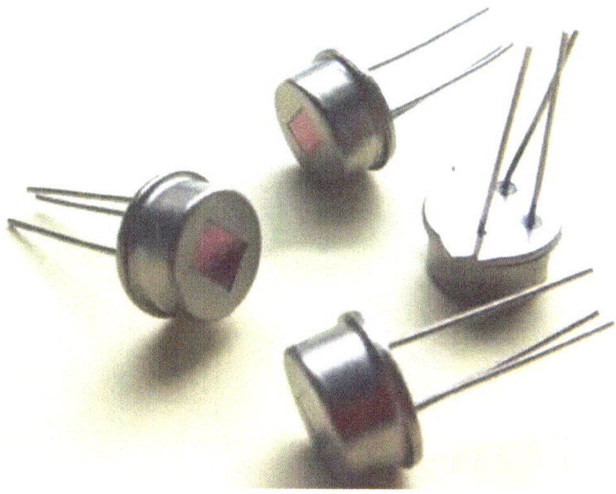

**Fig. 5.3** Pyroelectric infrared sensor

## 5.4 Related Technology of Infrared Sensor

### *5.4.1 Infrared Night Vision Technology*

Infrared night vision technology is divided into active infrared night vision technology and passive infrared night vision technology. The active infrared night vision technology is a night vision technology that performs observation by actively illuminating and utilizing the infrared light of the target reflection infrared source, and the corresponding equipment is an active infrared night vision apparatus. Passive infrared night vision technology is an infrared technology that realizes observation by means of infrared radiation emitted by the target itself. It finds the target according to the temperature difference or heat radiation difference between the target and the background or parts of the target. It is equipped with a thermal imager. The thermal imager has advantages different from the other night vision devices. For example, it can work in fog, rain and snow, and has a long range of action. It can recognize camouflage and anti-interference, and has become the development focus of foreign night vision equipment.

1. **Active Infrared Night Vision Technology (Near Infrared Region)**

The active infrared night vision system consists of an infrared emitter, a picture tube, a power supply, and an optical system. Its working principle follows the process of the following: infrared radiation is being emitted by the infrared emitter. When there is an obstacle detected ahead, the infrared radiation that touches the obstacle will be reflected which then falls onto the photoemitter of the infrared image tube through the objective lens group. The photoemitter then absorbs the photon energy to emit the photoelectrons. The photoelectrons are accelerated, enhanced, and imaged by the image tube, and finally bombarded onto the phosphor screen to form a visible light image. Finally, through the eyepiece zoom, one can observe the magnified image (Fig. 5.4).

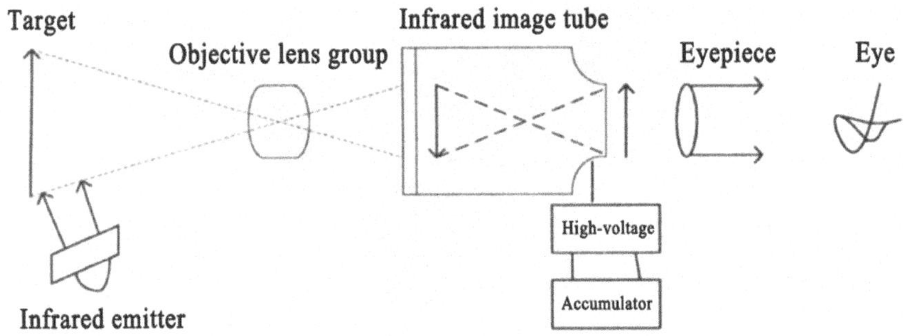


**Fig. 5.4** Active infrared night vision system

Its working band is in the near infrared region of 800–1000 nm. The range of the active infrared night vision system depends on the power of the infrared transmitting module. The greater the power, the farther is the range. However, due to the weight and volume limitations of the in-vehicle device, the actual infrared night vision system's range is also limited.

The third-generation infrared night vision technology is referred to as IR-III technology and belongs to an active infrared. Its principle is to emit infrared light with a wavelength of 850 nm through a PN junction emitting infrared light. According to the device for capturing images in the monitoring series, the CCD can sense the infrared spectral characteristics, perform active illumination, and reflect infrared light imaging. IR-III technology is a newly emerging infrared night vision technology that is the first-generation general LED that succeeds the second-generation LED Array. It is equipped with characteristics such as the ability of measuring night vision distance, zero light failure, low power consumption, high performance in photoelectric conversion and has a small size. The IR-III technology consists of only one light-emitting diode. It has a very low heat generation and its chip is only 50 °C. It was once called a "cold light source." Another feature of the IR-III technology is the use of the special packaging materials instead of the ordinary epoxy resin for the lens, which effectively prevents the internal lens material from damage (cracked lens surface and diminished light) due to the heat. This allows the continuous emission of the infrared without loss. Concentrating the beam at the focus of the lens also ensures the intensity of the light. In addition, IR-III technology has COB packaging technology, which can effectively reduce power consumption. The infrared light emitted by the single IR-III technology is 3.7 times that of the first generation of traditional LED infrared camera.

2. **Passive Infrared Night Vision Technology (Middle and Long-Range Infrared)**

In nature, all the objects that have a temperature that is higher than the absolute zero (−273 °C) will continuously radiate infrared rays. This infrared radiation carries the characteristic information of the object which provides an objective basis by using the infrared technology to distinguish among the various types of targets via the temperature level and heat distribution field.

Passive infrared night vision refers to the thermal imaging of the infrared night vision. The infrared radiation emitted by the object is concentrated on the photosensitive surface after spectral filtering and optical scanning of the optical system. The object is subsequently scanned by the horizontal and vertical scanner. The image of the object formed by the scanner is swept point by point on the detector. The detector then converts the received infrared light signal into a corresponding voltage signal. The voltage signal is then amplified by an amplifier and converted from digital-to-analog by the A/D converter which eventually forming an image signal to be displayed on the screen (Figs. 5.5 and 5.6).

### 5.4.2 *Infrared Binocular Stereo Vision*

The binocular stereo vision system uses two cameras to simultaneously acquire the images, and calculates the three-dimensional coordinates of the points by using the parallax of the corresponding points of the two images to achieve the three-dimensional positioning. Target recognition and point matching algorithms are the basis and focus of the three-dimensional positioning of the binocular vision. But within a certain range, these algorithms are very sensitive to noise and the use of artificial representations often changes the environment, and hence is inconve nient. In order to achieve reliable positioning of the target, many existing solutions use infrared cameras to perform binocular stereo vision or ordinary cameras with infrared filters instead of the infrared cameras to acquire specific infrared spectrum representation image.

Therefore, the infrared binocular stereo vision system night vision acquires three-dimensional information in the viewing space based on the parallax, that is, a

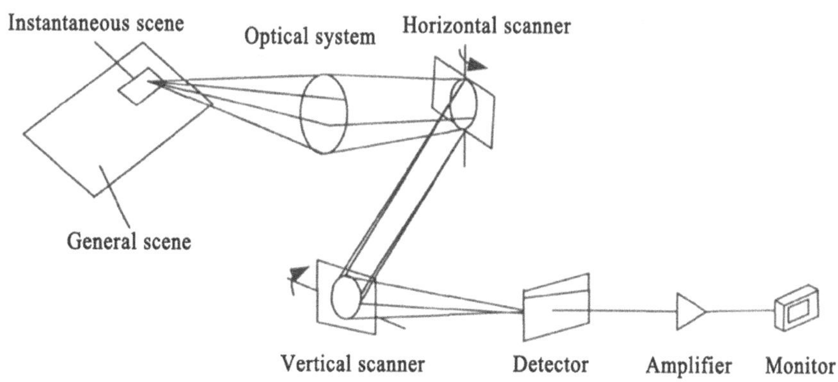


**Fig. 5.5** Passive infrared night vision system

**Fig. 5.6** Image of normal visible light and far range infrared

triangle is formed between the image plane of the two cameras and the front object, and the positional relationship between the two cameras is known. Based on the triangle similarity principle, the depth data, three-dimensional size of the object in the common field of vision of the two cameras and the three-dimensional coordinates of the feature points of the space object can be obtained. This method is not affected by the material of the object, and can enhance the operational reliability of the mobile platform including the drone and the manned helicopter.

If the infrared thermal imaging technology is integrated with the binocular stereo vision technology, it can realize the ability to penetrate through fog and cloud, immune to electromagnetic interference and operate at any time and under any weather without any auxiliary light source. In addition, wide dynamic imaging technique enables the imaging infrared camera to be able to adaptively adjust the image under sudden changes in temperature and hence ensures the real-time capturing of the obstacle range data (Fig. 5.7).

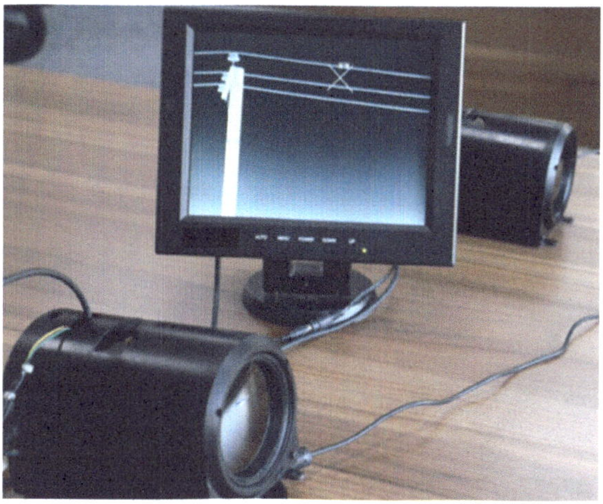

**Fig. 5.7** Infrared binocular stereo vision environment perception system

## 5.4.3 Pedestrian Detection

Pedestrian detection technology can be realized by using infrared stereo vision. The first stage is to use two different thresholds to focus on the input image area with high intensity values. As shown in Fig. 5.8a, these regions represent warm bodies. Initially, high threshold pixel values should be used to remove cold or hardly warm regions and select pixels that correspond to very warm bodies. If the pixel is adjacent to other selected pixels in a region-increasing manner, then pixels having gray levels that are higher than the lower threshold are selected. The resulting image contains only warm contiguous areas that represent hotspots, as shown in Fig. 5.8b. In order to select a vertical stripe that contains a hot area, column-by-column histogram needs to be calculated on the resulting image. Setting the value of the adaptive threshold is part of the average of the entire histogram. The histogram is filtered using an adaptive threshold. If a plurality of thermal objects is vertically aligned in an image, their contribution will be summed in the histogram. At the same time, based on the new directional histogram generated by computing the gray level of every stripe, similar horizontal stripe warm zones can be classified. After determining the warm zone, a rectangular bounding box frame is generated to mark the area where the pedestrians may be located, as shown in Fig. 5.9a. By refining the boundary box frame, the pedestrians can be detected accurately, as shown in Fig. 5.9b.

## 5.4.4 Target Tracking

The computer stereo vision includes a distance information system that is capable of acquiring, estimating, and extracting scenes in space from a set of 2D images. By

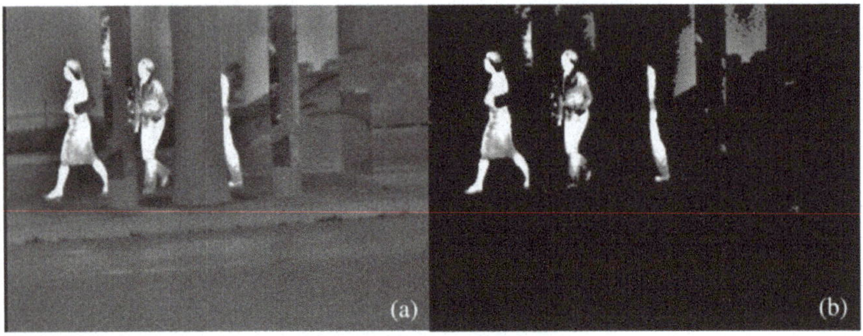


**Fig. 5.8** Pedestrian detection process

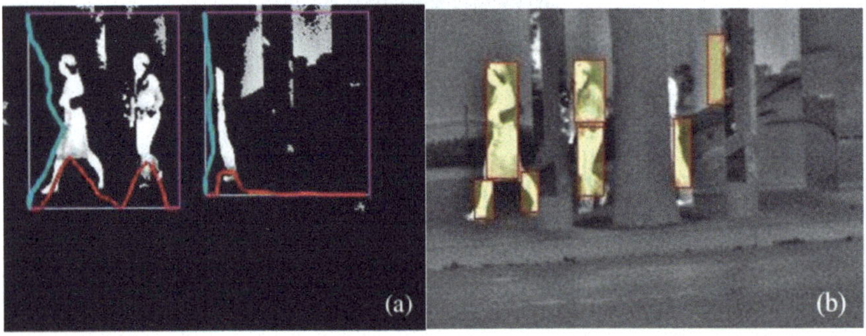


**Fig. 5.9** Warm zone element detection

using the long-wave infrared sensor, the system can be further upgraded to a system in low visibility scenario. The normal color sensor may not be the best choice. The data stream is converted to a serial USB for processing by the analog signal stream. The sensors are placed in parallel, mounted on a bracket and attached to the robot platform.

If two sensors are placed into a single scenario, the epipolar geometry could be used to calculate the distance based on the parallax between the matching feature points. The feature points are matched based on the cost from the sum of absolute difference (SAD) algorithm. A pixel block could be found in Fig. 5.10(left) which is then calculated for the SAD based on the pixel intensity of the target image (Fig. 5.10, right). The search is simplified by correcting two images. Each possible feature point is in the same pixel row and does not require a vertical search. A block with the least cost along each pixel row is considered the best match. This difference is inversely proportional to the actual distance of the object (person) and is then sent to the robot for tracking.

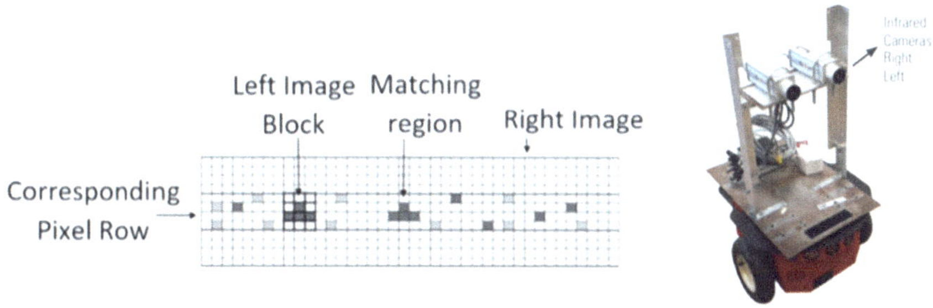


**Fig. 5.10** Target tracking structure and principle

# 5.5 Concept and Characteristics of Ultrasonic Sensors

## *5.5.1 Ultrasound*

Sound is transmitted in the form of sound waves. Acoustic waves are mechanical waves that are generated by the vibration of objects and propagate in all directions by means of various media. According to the frequency of sound waves, ultrasonic waves are generally referred to as sounds outside the frequency range of 20 Hz–20 kHz, while sounds below 20 Hz are called infrasound waves, and sounds above 20 kHz are called ultrasonic waves. The sound in the frequency range of 20 Hz–20 kHz is the sound that can be heard by the human ear. The frequency range of sounds spoken by ordinary people is 100 Hz–8 kHz, as shown in Fig. 5.11.

Ultrasonic waves propagate linearly. The higher the frequency, the weaker the diffraction ability, but the stronger the reflection capability. For this reason, an ultrasonic sensor can be developed by utilizing this property of the ultrasonic waves. Furthermore, ultrasonic wave propagation velocity in air is slower, about 340 m/s, which makes the use of an ultrasonic sensor very simple.

Ultrasonic waves have the following characteristics: ultrasonic waves can propagate in gases, liquids, solids, and living organisms, and in many cases where it can propagate in mediums where light, heat, electromagnetic waves, etc. cannot propagate. Ultrasonic wave is a type of elastic wave. Other than the longitudinal

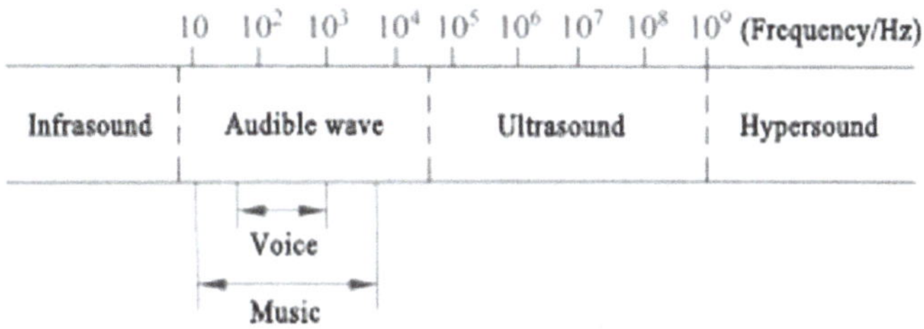


**Fig. 5.11** Sound wave frequency distribution

wave, there are also the transverse wave and surface wave. The longitudinal wave refers to the wave of which the vibration direction is consistent with the wave propagation direction; the transverse wave refers to the wave of which the vibration direction is perpendicular to the wave propagation direction; and the surface wave is the wave, of which the vibration of the particle between the transverse wave and the longitudinal wave propagates along the surface, that decays rapidly as the amplitude increases with depth. Due to the short wavelength, sharp directionality can be obtained for small sound sources, the energy is easily concentrated, and reflection and refraction are generated at interfaces with different acoustic impedances.

The wavelength of the ultrasonic wave is equal to the speed of sound/frequency, that is $\lambda = \frac{c}{f}$. The propagation speed varies in various media but is much lower than $3 \times 10^8$ m/s the propagation speed of the electromagnetic wave. When the wavelength is short, it is similar to ordinary light. For a small sound source, a sharp directional beam can be obtained, and the azimuth resolution is also high. If the ultrasonic wave is perpendicularly incident on the interface of two different acoustic impedances, the reflected wave and the transmitted wave are generated, and the reflectivity can be expressed as

$$Reflectivity = \frac{Reflected-Radiation}{Incident-Radiation}$$

The reflectivity between air and liquid, as well as air and solid, is 100%.

When sound waves propagate through the medium, the phenomenon of attenuation occurs successively which is attributed to two factors, namely physical and medium factors. In the former case, since the ultrasonic wave generally propagates as a spherical wave, the energy density decreases as the distance increases. For the latter, the wave energy becomes heat energy and is absorbed by the medium. In addition, when the acoustic impedance of the medium is different, scattering and reflected waves are generated. Therefore, the sound waves are successively attenuated. The higher the frequency, the greater the attenuation, and the larger the difference with the medium.

### 5.5.2 Ultrasonic Sensor Principle and Characteristics

The ultrasonic sensor is mainly composed of a transmitter, a receiver and a control portion. The transmitter and receiver complete the transmission and reception of ultrasonic waves, which are collectively referred to as ultrasonic probes or transducers, as shown in Fig. 5.12.

Classified according to the working principle, the transducer is divided into the piezoelectric type, the magneto strictive type and the electromagnetic type. Among them, the piezoelectric transducers are the most common, and the materials include mainly piezoelectric crystals and piezoelectric ceramics. At present, ferroelectric

**Fig. 5.12** Ultrasonic transducer

ceramics are the most widely used materials with piezoelectric effects. In addition, there are organic materials with piezoelectric effects, but due to their low stability, the applications to date are still very limited.

An ultrasonic transducer that emits sound waves into the air or that receives sound waves again from the air (different in use in liquids and solids) has a relatively large amplitude to release enough energy into the air or to receive the energy in the air again. The mechanical deformation of the piezoelectric ceramic itself is not sufficient to release enough energy, and hence the effect is mechanically strengthened. This can occur in some unmanned systems, such as the ultrasonic distance sensor of an unmanned vehicle, for which a piezoelectric ceramic foil is placed flat on the metal film. If an AC voltage is applied between the electrodes, the diameter and thickness will change (as shown in Fig. 5.13). Since the sheet is attached to the metal film, the change is transmitted in accordance with the bending vibration of the film driven by the string frequency, which produces a large vibration amplitude.

Conversely, the colliding sound waves cause the diaphragm to produce bending vibrations and thereby cause a change in the diameter of the piezoelectric ceramic sheets. This will produce an AC voltage between the electrodes, which is then electrically amplified and reprocessed. In most cases, ultrasonic accumulators are used not only for transmitting but also for receiving ultrasonic waves.

The vibrating diaphragm must be fixed on its edge. In practice, it is achieved by attaching a piezoelectric sheet to the bottom of a small aluminum can. The bottom acts on the diaphragm, and the strong side walls of the small can hold the outside of

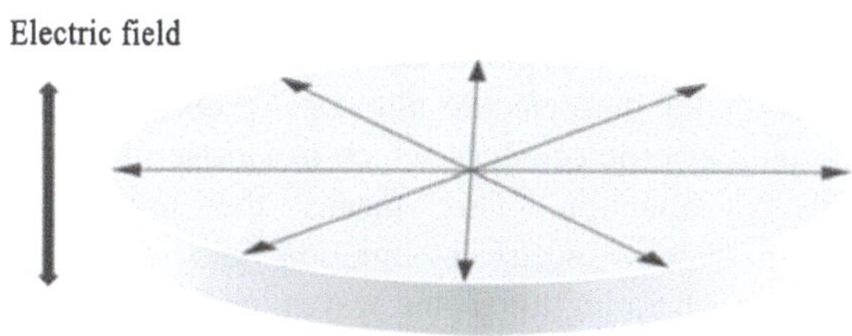


**Fig. 5.13** Planar vibration of piezoelectric ceramic sheets

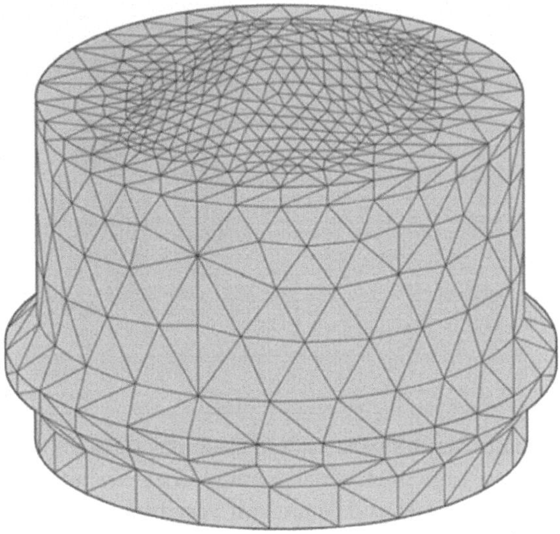

**Fig. 5.14** Finite element analysis of vibrating film

the diaphragm. The vibration is mainly concentrated on the diaphragm, but the side walls also participate in the movement to a lesser extent. This is important because the tension of such a small aluminum can affect the movement of the diaphragm (Fig. 5.14).

The diaphragm needs to be silenced in order to make the accumulator more robust and easier to control. For instance, using materials and structures such as the silicone foam that matches the operating frequency. Although this will reduce the efficiency of the accumulator, there are several advantages such as the ability to instantly absorb harmful sound waves that enter the interior of the sensor, and increased robustness of the diaphragm against temperature and aging effects caused by external pollutants or moisture and frequency change.

## 5.6 Structure and Type of Ultrasonic Sensors

### *5.6.1 Basic Structure of Ultrasonic Sensor*

Ultrasonic sensors, which are also known as an ultrasonic probe, can be classified based on its operating principle into piezoelectric, magneto strictive, electromagnetic, etc., of which piezoelectric is the most commonly used. The commonly used materials for piezoelectric ultrasonic probes include the piezoelectric crystals and the piezoelectric ceramics which make use of the piezoelectric effect of the materials to function. This means the inverse piezoelectric effect converts high-frequency electrical vibration into high-frequency mechanical vibration, thereby generating ultrasonic waves, and can be used as a transmitting probe. The positive piezoelectric effect converts ultrasonic vibration waves into electrical signals, which can be used as a receiving probe. Due to the reversibility of the piezoelectric effect, some ultrasonic sensors in practical applications use a single probe for both the ultrasonic transmission and reception.

Figure 5.15 shows the basic structure of the ultrasonic probe. It is mainly composed of piezoelectric wafer, absorption block (damping block), protective film, metal shell, lug and other components. The piezoelectric wafer is mostly in the shape of a disk, and the ultrasonic frequency is inversely proportional to its thickness. Both sides are plated with a silver layer as a conductive plate. Damping blocks are used to reduce the mechanical quality of the wafer and absorb sonic energy. If there is no damping block, when the energized electrical pulse signal stops, the wafer will continue to oscillate, lengthening the pulse width of the ultrasonic wave, making the resolution of the ultrasonic sensor worse.

## 5.6.2 Type of Ultrasonic Sensor

Ultrasonic probes come in many different configurations and can be classified into straight probes (longitudinal waves), oblique probes (transverse waves), surface wave probes (surface waves), lamb wave probes (Lambo waves), dual probes (one reflection probe and one receiving probe) and so on. Figures 5.16 and 5.17 shows the structure of several ultrasonic probes and the physical map of several different configurations of ultrasonic probes.

**Fig. 5.15** Structure of the ultrasonic probe

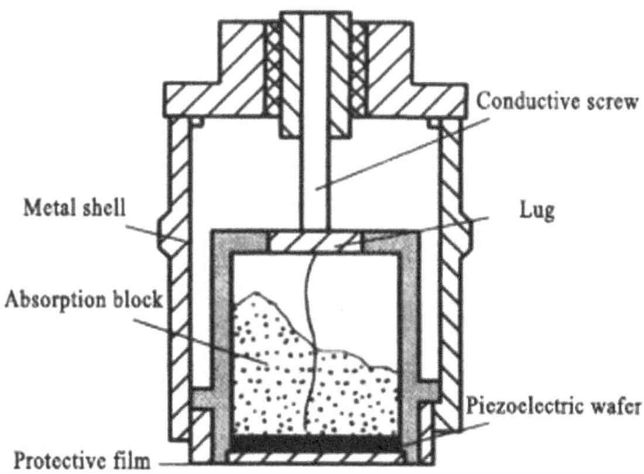

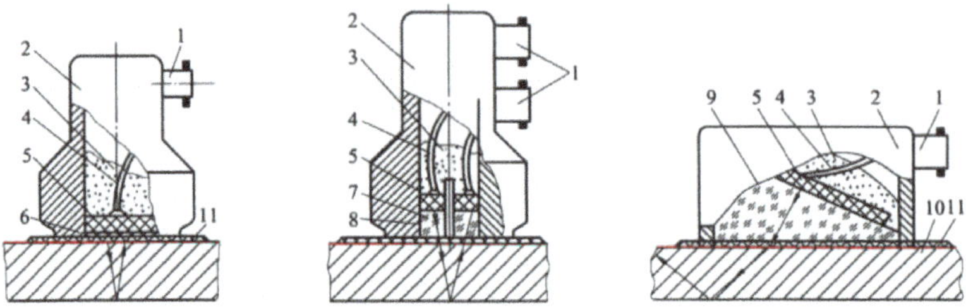

(a) Single Crystal Straight Probe (b) Double Crystal Straight Probe (c) Oblique Probe

1—Connector; 2—Outer Casing; 3—Damping Absorption Block; 4—Lead; 5—Piezoelectric Crystal;

6—Protective Film; 7—Isolation Layer; 8—Delay Block; 9—Plexe Glass Wedge; 10—Test Piece; 11—Coupling Agent

**Fig. 5.16** Structure of several ultrasonic probes

(a)Single Crystal Straight Probe (b) Double Crystal Straight Probe (c)Oblique Probe

**Fig. 5.17** Diagrams of different structured ultrasonic probes

## 5.7 Ultrasonic Sensor Technology

### *5.7.1 Anti-jamming Technology of Ultrasonic Sensor*

In an unmanned system, such as for the vehicle reversing radar system in an unmanned vehicle, the ultrasonic sensor will transmit 40 kHz ultrasonic waves and receive 40 kHz ultrasonic signals in return. In reality, there exists signals of the same frequency or multi-frequency transmission in the environmental space. This could lead to misjudgment if not handled properly which in turn could affect the normal operation of the system and hence reduce the reliability of the system. Therefore, in addition to hardware filtering, software filtering is usually performed to remove interference signals which greatly reduces the false positives.

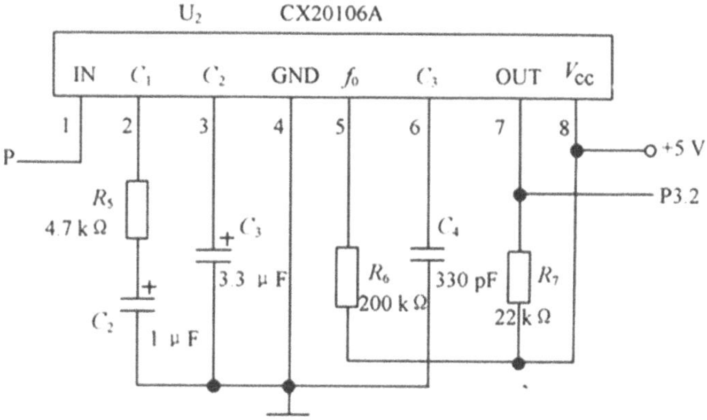


**Fig. 5.18** Ultrasonic ranging application circuit

The following figure illustrates an example of the ultrasonic sensor hardware design of which the ultrasonic electric signal amplifying circuit uses the integrated circuit CX20106A. By setting the center frequency of the internal band-pass filter circuit, $f_0$, to 40 kHz through the externally connected resistor, the amplified ultrasonic signal could be received and negative pulse voltage will be output.

The circuit diagram of the ultrasonic ranging application is shown in Fig. 5.18. The denoted pin 1 is the ultrasonic electric signal input end, and the RC serial network is connected between the pin 2 and the ground, which is an integral part of the internal preamplifier circuit negative feedback network. The value of resistor R5 determines the gain of the preamplifier circuit. The R5 resistance value decreases, the negative feedback decreases, and the amplification factor increases. Conversely, the amplification factor will decrease. The detection capacitor C3 is connected between the pin 3 and the ground which facilitates the size of the capacitor C3 to be appropriately changed. This allows the sensitivity and anti-interference ability of the ultrasonic electric signal amplification and shaping circuit can be changed. C3 has a large capacitance, low sensitivity and strong anti-interference ability. C3 has small capacity, high sensitivity, weak anti-interference ability, and easy to cause malfunction. A resistor is connected between the pin 5 and the power supply to set the center frequency $f_0$ of the internal band pass filter circuit. When R6 = 200 kΩ, $f_0$ = 40 kHz. An integral capacitor is connected between pin 6 and the ground. The standard value is 330 pF. If the capacitance value is too large, the detection distance will be shortened. Pin 7 is the open collector output of the circuit. R7 is the pull-up resistor for this pin. When the integrated circuit CX20106A has no signal input, pin 7 output is high. When the input ultrasonic electric signal is amplified and shaped, the 7-pin outputs a negative pulse voltage.

The software design of the ultrasonic sensor uses the arithmetic average filter in the digital filtering program to repeatedly measure for each detecting point. The reliability of the data collection is increased by taking the average of the detected ranging points as the measured data. To minimize the detection dead zone, the set delay time can be determined according to the residual vibration time of the ultrasonic sensor used. The minimum delay time can be determined in actual commissioning.

The working environment is very poor when the reversing radar is installed in the car. When the car is reversing, the high-voltage ignition generates strong electromagnetic radiation which will affect the normal operation of the circuit. Therefore, from the hardware and software aspects, should consider taking anti-interference measures to improve the reliability of the system work. For instance, using metal shell to shield the circuit and using shielded wire to connect the ultrasonic sensor. Under the situation when the conditions of measuring the range is satisfied, the capacity of the ultrasonic capacitor that is used for amplifying and shaping the C3 capacitor in the shaping circuit can be appropriately adjusted. By adding a surveillance circuit to the hardware, setting up software traps of redundant instructions to the software design or a surveillance software could prevent the program from entering an indefinite loop. To the driver, the main concern when reversing is on the presence of the obstacles at the rear of the car and the distance between the obstacle and the car. Since there is inertia when the car is braking, the driver always brakes in advance when the vehicle encounters an obstacle. Considering the cost performance, the accuracy of the reversing radar measurement does not have to be high. However, from the perspective of safety, the measurement value is preferred to be big rather than small.

## 5.7.2 Sector Scanning Detection of the Ultrasonic Sensor

Ultrasonic sensor ranging has a more obvious drawback than the other radar sensors, that is, it is impossible to accurately describe the position of the obstacle. The ultrasonic radar returns a value of the detection distance as shown in Fig. 5.19. The obstacles at A and B will return the same detection distance d. Therefore, knowing only the detection distance, d, via a single radar cannot determine whether the obstacle is located at A or at B.

In the same environment, as the basis of the ultrasonic ranging system only consists of the echo time, so the ultrasonic sensor becomes the center in the effective detection range of the system. Objects of the same arc will produce the same echo time and receive the same detection results. In other words, the distance measured by using the ultrasonic ranging system is not necessarily the distance of the object directly in front of the sensor. Ultrasonic ranging can only measure the distance of the measured object, but it is impossible to determine the exact direction of the object that produces the distance. Furthermore, it is impossible to determine whether there is only one object being measured. The larger the sensor beam angle selected by the ranging system, the more uncertain is the specific orientation of the measured object which will result in the directivity of the ranging to be worsen. Insufficient directivity is the biggest disadvantage of ultrasonic ranging.

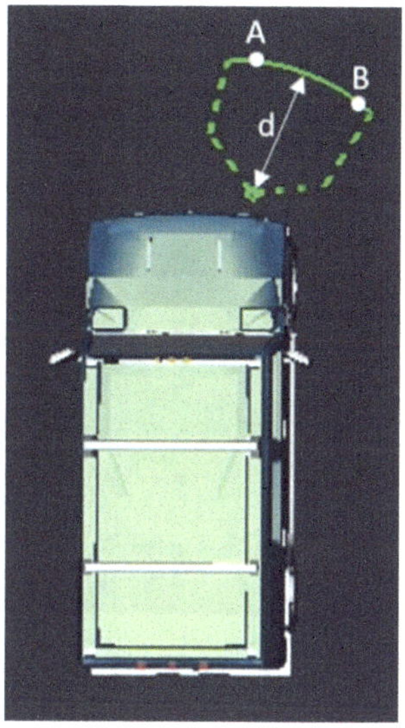


**Fig. 5.19** Ultrasonic radar ranging

In the previous design of the ultrasonic radar, numerous ultrasonic sensors were utilized based on the uniform array detection method. Additionally, it requires the sum of the detection ranges of all the sensors to fully cover the areas near the vehicle body to ensure full detection of the parked environment.

Since the directivity of the ultrasonic sensor is inversely proportional to the detection range, if the system is required to more accurately measure the position of the obstacle, a dense array of small beam sensors must be used. However, such an approach is costly and affects aesthetics. More importantly, if the dense array of sensors may cause interference with each other, the reliability of the detection is affected. Therefore, most of the current UAV ultrasonic radar systems sacrifice the accuracy of the measurement, and select 3–4 large beam angle ultrasonic sensor arrays with large detection range.

Such a design method can only detect the existence of an obstacle, but cannot indicate the specific orientation of the obstacle. Moreover, such a detection method still has a certain dead zone. As shown in Fig. 5.20, if the smaller obstacle is located between the two sensors at a position closer to the rear of the vehicle, it is likely to be missed by the system. This is due to the characteristics of the ultrasonic ranging system itself and is difficult to overcome.

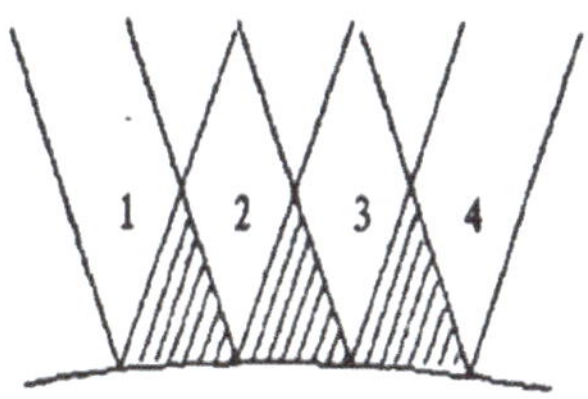


**Fig. 5.20** Ultrasonic sensor detection dead zone

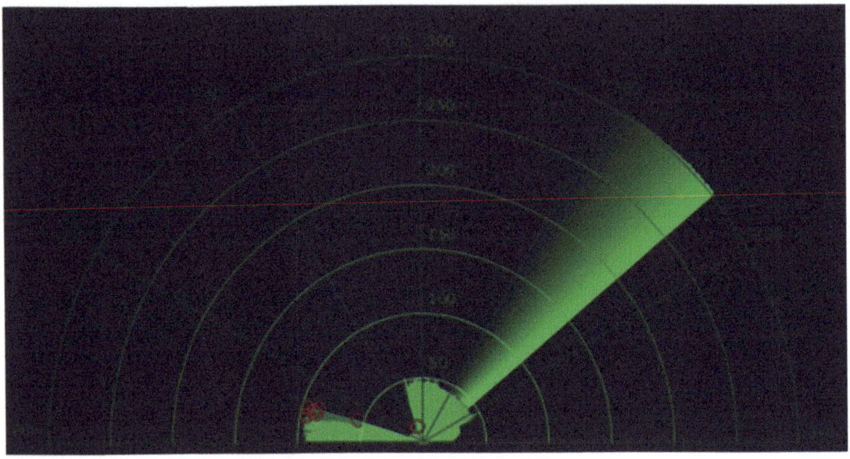

**Fig. 5.21** Sector scan

The labelled area in the figure is the effective detection range for each sensor, and the shadow part is the detection dead zone between the adjacent sensor detection areas.

In order to overcome the shortcomings of the ultrasonic sensor's poor directivity in ranging and the presence of blind zones, the ultrasonic sensor technology that is currently used by the unmanned systems adopts the sector scanning detection method. In this design, a stepper motor with a step angle of 7.2° is used to drive an ultrasonic sensor with a beam angle of 5°. In each round of scanning, the motor performs 20 steps and scans a range of 144° in front of the vehicle. By doing so, from the starting position, the ultrasonic sensor performs ranging in a total of 21 different angles. Every time the stepping motor steps at an angle, the ultrasonic sensor would measure the distance signal at the same angle. Combined with the current scanning angle, a more accurate position that includes information such as the distance and direction would be obtained. Based on this information, it is possible to determine the specific orientation of the obstacle more accurately and to obtain relatively accurate environmental information around the vehicle (Fig. 5.21).

## 5.8 Application of Ultrasonic Sensor in Unmanned System

### *5.8.1 Application of Ultrasonic Sensor in Unmanned Vehicle*

As afore-mentioned, the ultrasonic sensors enjoy a greater advantage for short distance and low speed measurement. Hence, the ultrasonic sensors can aid the vehicles that are stopping at low speed to detect the surrounding objects in the unmanned vehicle system. Ultrasonic parking assistance is also known as the parking assist systems, the parking guidance systems, and the reverse assist. These systems are able to facilitate from simple detection of the surrounding objects and then alert the driver by generating a sound to automatic parking with virtually no human intervention. Typically, these systems have 4–16 sensors that are cleverly mounted around the body to provide the required inspection coverage.

There are two common types of ultrasonic radar. The first type refers to those installed on the front and rear bumpers of the car, that is, the reversing radar which is used to measure the obstacles in front and at the rear of the car. This type of radar is called UPA in the industry. The second type refers to radars that are installed on the sides of the car and are used to measure the distance of the side obstacles, also known as APA in the industry. The detection range of UPA ultrasonic radar is generally between 15 and 250 cm, which is mainly used to measure for obstacles in front and at the rear of the car. The detection range of APA ultrasonic radar is generally between 30 and 500 cm. As the APA's detection range is farther, so it is more expensive and powerful than the UPA. APA's advantage in the detection range could not only facilitate the detection of the obstacles on the left and right side of the car, but also determine the availability of a car parking lot based on the data received from the ultrasonic radar's returned signal. As illustrated in Fig. 5.22, a total of 8 UPA have been installed at the front and rear of the car, and a total of 4 APA have been installed on the left and right side of the car. It can be clearly seen that the detection range and area of the UPA and APA are significantly different.

A common requirement of the ultrasonic parking assist module includes the ability to detect objects from 30 cm to 5 m. In order to satisfy the requirements of the self-driving vehicle, the short- and long-distance object detection standards have

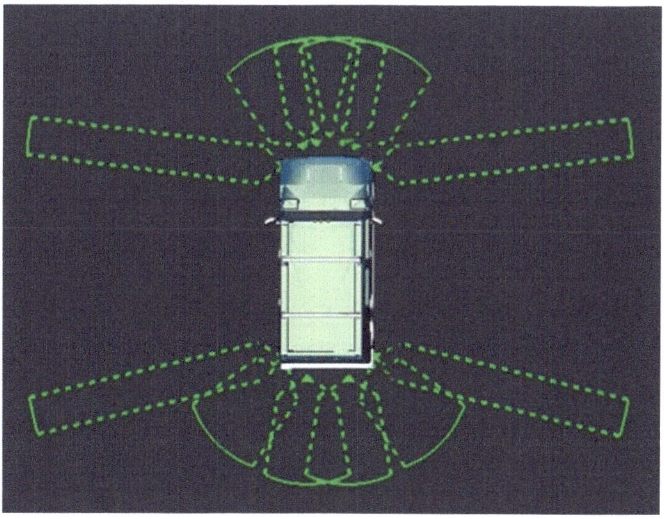

**Fig. 5.22** Diagram of the ultrasonic radar configuration in an unmanned vehicle

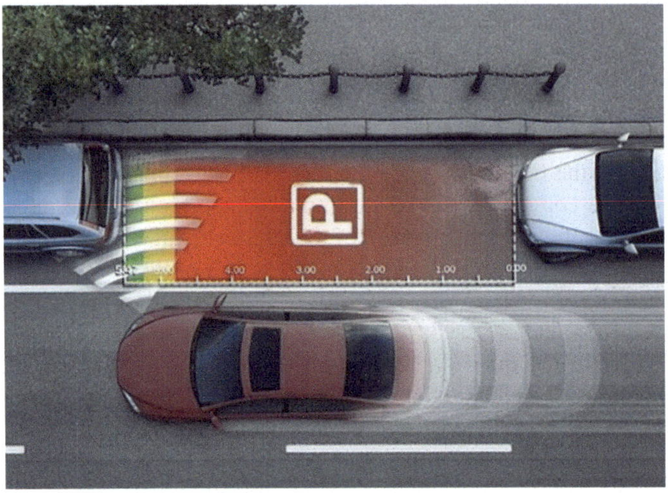


**Fig. 5.23** Ultrasonic sensor parking assist

become more stringent. Starting from 2025, the ultrasonic module will have to be able to detect objects between 10 cm and 7 m (Fig. 5.23).

At present, automakers have produced vehicular models that could support the use of the car key to remotely control the car for automatic parking. During the operation, the user only needs to indicate two commands, namely forward or backward, and the car will continue to use the ultrasonic sensors to detect for parking space and obstacles. Base on this information, the car will automatically steer the wheel or brake accordingly to realize the automatic parking function. Among which, one of the current automakers which uses the third-generation ultrasonic semi-automatic parking systems adopts the parking assist system which usually uses 6–12 ultrasonic radars. The 4 short-range ultrasonic radars at the rear of the car are responsible for detecting the distance between the reversing car and the obstacles while the long-range ultrasonic radar is responsible to detect for parking space. Engineers have continuously been mining the potential of the ultrasonic radar, given that it is the lowest cost sensor in an unmanned vehicle.

### 5.8.2 Application of Ultrasonic Sensor in UAV

In recent years, consumer drones have become increasingly popular for various uses such as to capture breath-taking views, to deliver relief aid and for competition. Most of the UAV use a variety of the sensing technologies for autonomous navigation, collision detection, and many other functions. The ultrasonic sensor is especially helpful in UAV landing, hovering and ground tracking. The UAV landing aid is a function of the UAV that can detect the distance between the bottom of the drone and the landing area, determine whether the landing point is safe, and then slowly descend to the landing area. Although GPS monitoring, barometric sensing and other sensing technologies contribute to the landing process, in this process, ultrasonic sensing is the primary and most accurate basis for the UAV. Most of the UAV also possess hovering and ground tracking modes that are primarily used to capture footage and land navigation. The ultrasonic sensors can also help to keep the UAV at a constant height above the ground.

Similar to many ultrasonic sensing applications, the UAV landing assistance system adopts the concept of the time of flight (ToF). ToF is estimated as the total time taken for the ultrasonic wave that is emitted from the sensor to reach the target object and then reflected from the object back to the sensor, as illustrated in Fig. 5.24.

At point 1 in the figure, the ultrasonic sensor of the UAV emits sound waves, which are represented as saturated data on the return signal processing path. After transmission, the signal processing path becomes muted (point 2) until the echo is reflected back from the object (point 3).

While numerous sensing technologies can detect the proximity of the objects, the ultrasonic sensing works exceptionally well in terms of detection distance, solution cost, and the reliability on different surfaces when the drone is landing. A common requirement for UAV ground tracking and landing is the ability to reliably detect distances at 5 meters above the ground. Assuming that the signal is properly tuned and processed, ultrasonic sensors in the 40–60 kHz range typically meet this range.

The ultrasonic sensor is able to detect on surfaces that the other technologies are unable to resolve. For instance, drones often encounter glazing and other glass surfaces on the buildings. Light sensing technology sometimes passes through glass and other transparent materials, which makes it difficult for drones to hover over glass buildings. Ultrasonic waves, however, can reliably reflect on the glass surface. Although the ultrasonic sensor is mainly used for UAV landing aids and hover, its

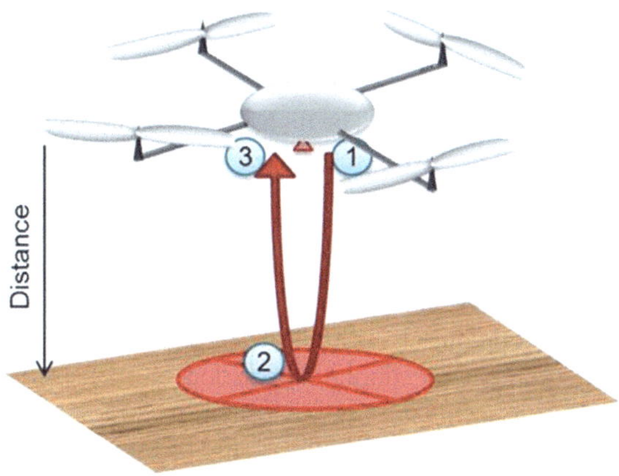


**Fig. 5.24** ToF of ultrasonic wave during UAV landing

**Fig. 5.25** Application of UAV ultrasonic radar

high cost-effectiveness is prompting UAV designers to explore other applications of this technology. There is great potential in the rapid development of the UAV (Fig. 5.25).

### *5.8.3 Application of Ultrasonic Sensor in Unmanned Boat*

With the continuous development of science and technology, in the civil and military fields, products are trending towards unmanned and intelligent development. In reality, the unmanned boats are essential in activities such as the environmental monitoring of the rivers that is not conducive for manual environmental surveillance, the monitoring of the leakage of nuclear radioactive materials and the sightseeing boats at the water city. Once the unmanned boat is equipped with the advanced control systems, sensor systems and communication systems, it can perform a variety of war and non-war military missions such as investigate, search, detection and monitoring.

At present, many existing unmanned surface crafts (USVs) have the function of intelligent navigation. The high-precision and high-cost radar and sensors make the products expensive, which is not conducive for large-scale application in the civilian field. The low cost and high cost-effectiveness are the advantages of the ultrasonic sensors. One of the USVs uses three ultrasonic probes with 270° field of view to detect the obstacles, and thus ensure efficient autonomous navigation. In the course of navigation, the electronic compass is used to monitor the hull angle as a closed-loop control system. The feedback input ensures the reliability and robustness of the USV during navigation and obstacle avoidance. The GPS/Beidou navigation technology and RF data wireless transmission technology are used to monitor the USV navigation status in real-time and to draw a motion trajectory map. When the obstacle avoidance is successful, the route optimization to the

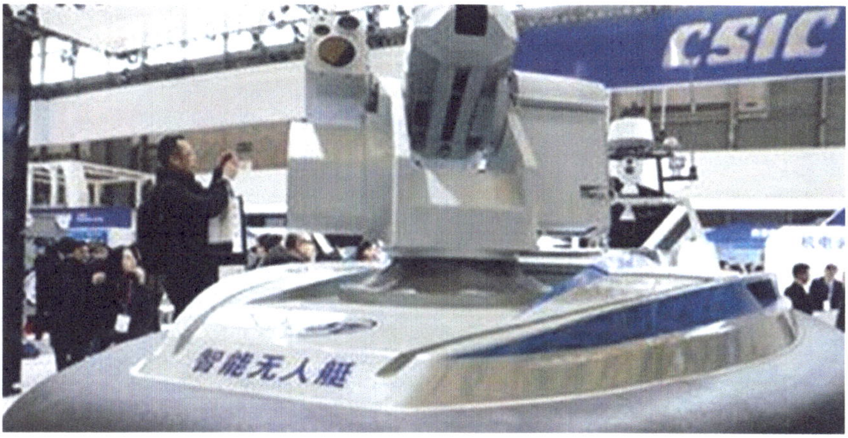


**Fig. 5.26** Ultrasonic radar-based surface intelligent unmanned boat

destination is conducted to reduce power consumption. The ultrasonic detection module estimates the distance between the USV and the obstacle based on the emitted and reflected sound waves to achieve the obstacle avoidance. In order to avoid the direct interference of the sound waves, the 3 ultrasonic probes adopted the cycle scanning method which ensures that only one probe is in operation at a time.

At the same time, in addition to the general obstacle avoidance function, the ultrasonic sensor can also perform image stabilization of a large field of view and three-dimensional combined obstacle avoidance in a complex background, further improving the performance of the unmanned boat, so as to achieve accuracy in navigation, successful avoidance and clarity in sight. The ultrasonic sensor becomes more attractive in unmanned boat applications with its excellent performance and low cost (Fig. 5.26).

## Bibliography

1. Wen L, Jin X, Lu X (2006) The development of infrared detection technology. Laser Infrared 36(9):840 [world record]
2. Wang D, Liang H, Qiu N, Xu S (2007) Application and analysis of infrared detection technique. Infrared Laser Eng 36(9):107[recorded]
3. Zhang X, Liang X (2013) Development requirements of infrared detectors. Electro-optic Control 20(2):41
4. Chen YB (2008). Pyroelectric infrared sensor principles and applications. Instrum Technol 6:66 [Worry about first]
5. Xiao J, Yu Z, Sheng Z (2010) Research progress of infrared array detectors. Commodity Quality 10:91
6. Li Q (2004) Pyroelectric infrared sensor. Age Electr Appliances 9:63
7. Meng X, Song B, Xu L (2007) Pyroelectric infrared sensor and its typical application. Instrumentation Users 14(4):42

8. Li F, Dong M, Xu R et al (2011) Research progress of pyroelectric infrared sensor in the field of biometric identification. Mod Instrum 3:10
9. He B (2014) Research on CO gas concentration detection system based on infrared spectrum. China Institute of Metrology
10. Xiao X (2011). System gas concentration measurement based on the infrared sensor. Tianjin University [heart through]
11. Bertozzi M, Broggi A, Lasagni A et al (2005) Infrared stereo vision-based pedestrian detection. In: IEEE proceedings. Intelligent vehicles symposium. IEEE, pp 24–29
12. Kong W, Zhang DB, Wang X et al (2013) Autonomous landing of an UAV with a ground-based actuated infrared stereo vision system. In: IEEE/RSJ international conference on intelligent robots and systems. IEEE, pp 2963–2970
13. Mohd MNH, Kashima M, Sato K et al (2014) A non-invasive facial visual-infrared stereo vision based measurement as an alternative for physiological measurement. In: Computer vision—ACCV 2014 Workshops. Springer International Publishing, pp 684–697
14. Sadeghipoor Z, Thomas JB, Süsstrunk S (2016) Demultiplexing visible and near-infrared information in single-sensor multispectral imaging. Color Imaging Conf 2016(1):76–81
15. Wu L (2015) Research on infrared binocular stereo vision significant target ranging technology [D]. Nanjing University of Science and Technology
16. Shi J (2017) Optimization of vehicle-assisted driving system based on far infrared technology. Inf Technol 12:70–72
17. Tong Y (2016) Research on key technologies of stereo vision based on infrared and visible dual-band images [D]. Tianjin University
18. Fan J (2016) Research on infrared vehicle safety technology [D]. University of Electronic Science and Technology

# Chapter 6
# Multimodal Sensor Collaborative Information Sensing Technology

## 6.1 Introduction

To-date, any single function sensor is unable to guarantee the provision of completely reliable information at any time or place. Therefore, the integrated consideration of the various sensors' advantages, the comprehensive use of the multiple sensors' data redundancy to complement one another's characteristics, and the collecting and acquiring of the multiple sensors' data for organic synthesis, that is, by using multi-sensor data fusion technology, the integrated information required for system operation could be obtained. This has become the key research of the unmanned system and is a problem to be solved. Multi-sensor information fusion is actually a functional simulation of the complex problem of human brain synthesis. Compared with a single sensor, in the aspect of solving problems such as detection, tracking and target recognition, the multi-sensor information fusion technology can enhance the survivability of the system, improve the reliability and robustness of the whole system, enhance the credibility of the data, improve the accuracy, expand the time and space coverage of the system, and increase the real-time performance of the system and information utilization. The characteristics of data fusion are as follows:

(1) Redundancy: Pertaining to parts with same attribute, the data of different sensors can be combined to improve the fault tolerance and robustness of the sensor. For example, two superimposed images.
(2) Complementary: Data fusion from different azimuth sensors to detect attributes of different parts. For example, using four camera probes to detect a 360° range.
(3) Cooperative: Pertaining to a part of same attribute, using a single sensor is unable to satisfy the detection requirements. By fusing data from different sensors, the surroundings can be better perceived. For example, stereo vision.



X. Bi, *Environmental Perception Technology for Unmanned Systems*, Unmanned System Technologies, https://doi.org/10.1007/978-981-15-8093-2_6

## 6.2 Levels and Classification of Data Fusion

### *6.2.1 Level of Data Fusion*

#### 6.2.1.1 Centralized Architecture

The extreme case of centralized processing refers to where all data processing and decision making are done in the same location, and the data is "raw data" from different sensors.

**Advantages:**

Sensor Modules—Sensor modules are small, low cost, and low in power consumption because they only need to perform inspection and data transfer tasks. The mounting position of the sensor is also flexible and requires minimal installation space. The replacement cost is low. Typically, sensor modules have lower functional safety requirements due to no processing or decision making.

ECU Processor—The central processing ECU can acquire all data because the data is not lost due to pre-processing or compression within the sensor module. Due to the lower cost of the sensor and the smaller form factor, more sensors can be deployed.

**Disadvantages:**

Sensor Modules—Processing sensor data in real time requires broadband communication (up to several Gb/s), so high electromagnetic interference (EMI) can occur.

ECU Processor—The central ECU needs high processing power and speed to handle all input data. For many high-bandwidth I/O and high-end application processors, this means higher power requirements and greater heat dissipation. The increase in the number of sensors will greatly increase the need for central ECU performance. By using an interface such as the FPD-Link III, the sensor's power consumption, control and configuration of multiple data (bidirectional reverse channel) is transmitted on a coaxial cable (Figs. 6.1 and 6.2).

#### 6.2.1.2 Fully Distributed Architecture

The fully distributed architecture is performed by the local sensor module for advanced data processing and decision making to a certain extent. The fully distributed system only sends object data or metadata (data describing the object characteristics and/or identifying the object) back to the central fusion ECU. The ECU combines the data and ultimately decides how to perform or react.

**Advantages:**

Sensor Module—A lower bandwidth, simpler and less expensive interface can be used between the sensor module and the central ECU. In many cases, a CAN bus of less than 1 Mb/s is sufficient.

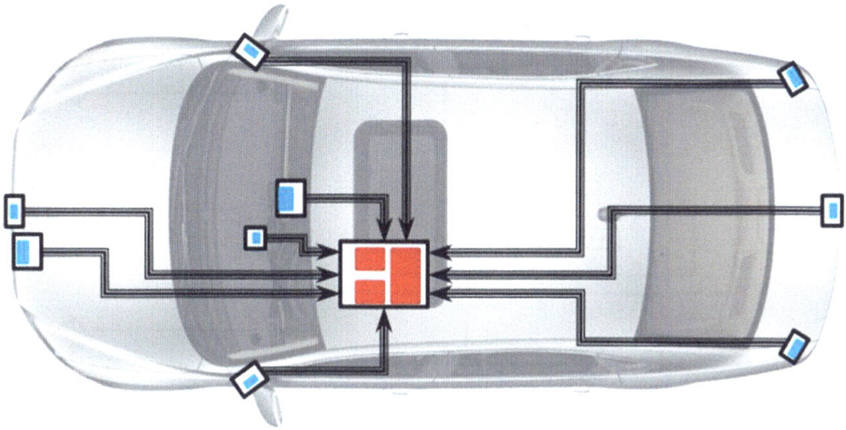

**Fig. 6.1** Diagram of the centralized architecture

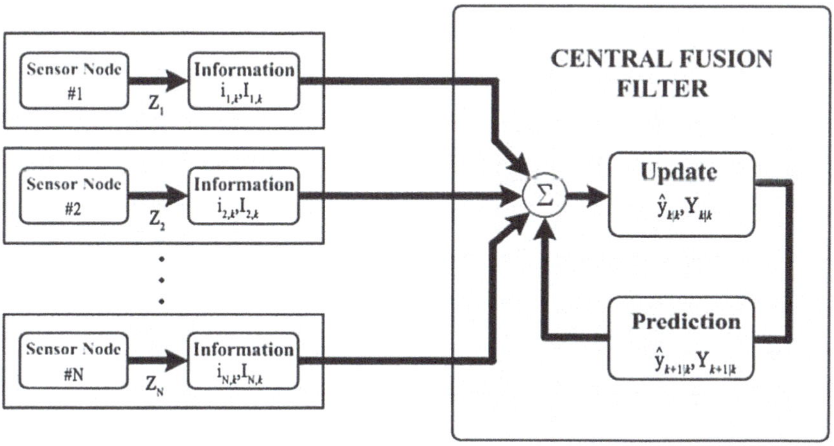


**Fig. 6.2** Concept of the centralized architecture

ECU Processor—The central ECU only fuses the object data together, so it requires less processing power. For some systems, an advanced secure microcontroller is sufficient. The module is smaller and requires less power. Since much of the processing is done inside the sensor, the increase in the number of sensors does not significantly increase the performance requirements for the central ECU.

**Disadvantages:**

Sensor Modules—The sensor modules require an application processor, which makes them larger, more expensive, and more power consuming. Due to the local processing and decision making, the functional safety requirements of the sensor modules are higher. With more sensors, the cost will increase dramatically.

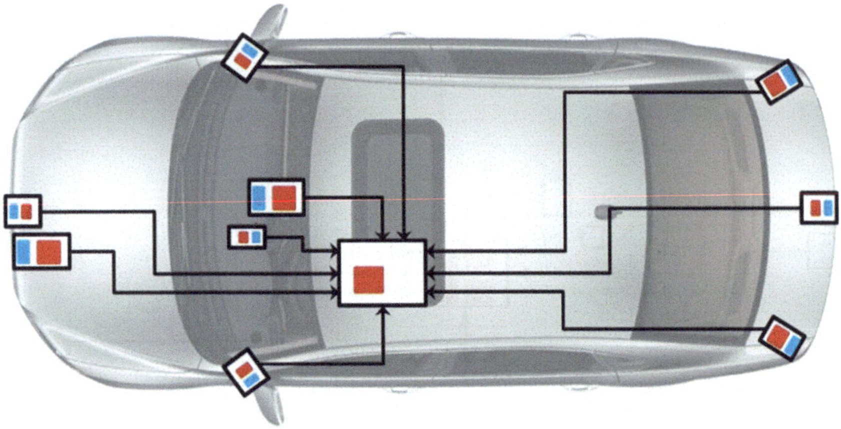

**Fig. 6.3** Diagram of a fully distributed architecture

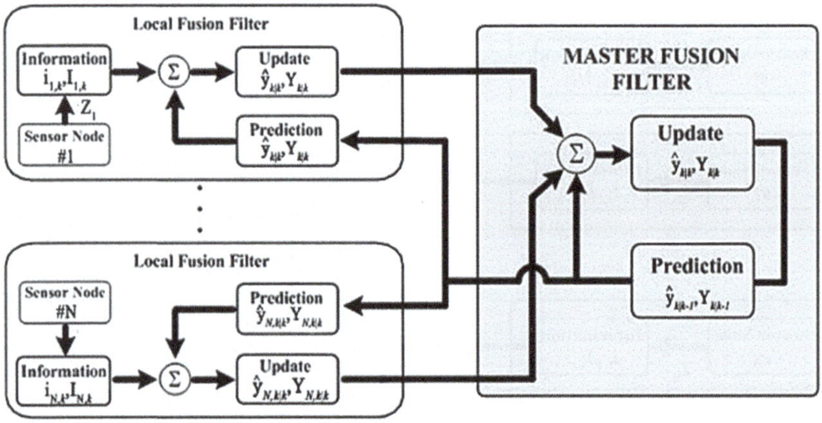


**Fig. 6.4** Concept of the distributed architecture

ECU Processor—The central decision-making ECU can only acquire object data and cannot access actual sensor data. Therefore, it is difficult to "enlarge" the area of interest (Figs. 6.3 and 6.4).

#### 6.2.1.3 Hybrid Architecture

Depending on the number and type of sensors used in the system, as well as the scalability requirements for different models and upgrade options, an optimized solution can be achieved by mixing the two topologies together. Many fusion

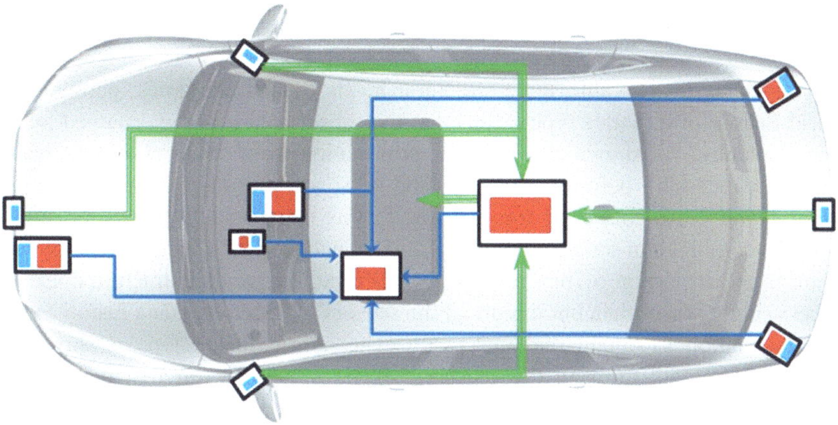

**Fig. 6.5** Diagram of the hybrid architecture

systems currently use sensors with local processing for radar, lidar and front cameras for machine vision. The hybrid architecture includes both the raw data and pre-processed data. It retains the advantages of the afore-mentioned systems. However, a relatively high price needs to be paid in the aspects of communication and computation (Figs. 6.5 and 6.6).

**Fig. 6.6** Concept of the hybrid architecture

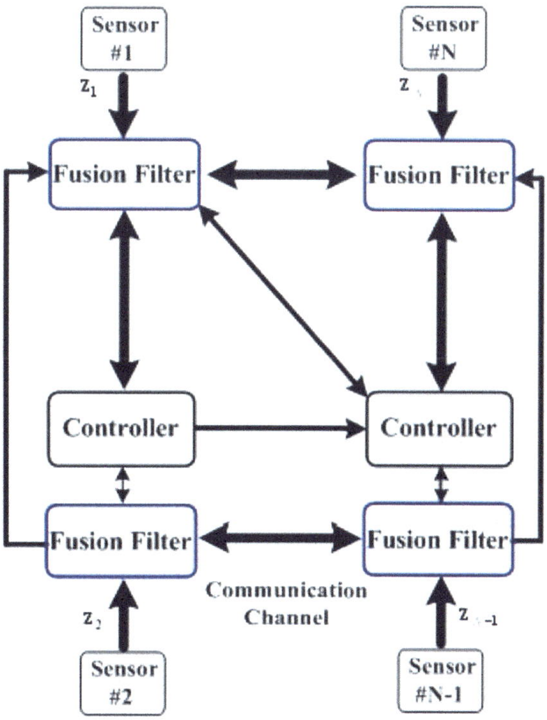

## 6.2.2 Types of Data Fusion

Data fusion is divided into three structural levels, which are series, parallel and hybrid fusion form. Series type means that the sensor data is being arranged into a series structure. As shown in Fig. 6.7a, information from sensor 1 is first being input for processing before fusing with the sensor 2 information. The data from sensor 1 and 2 are combined for analysis. In this manner, all the sensors' information is combined in series to obtain the final conclusion. The parallel type of data fusion is shown in Fig. 6.7b. Unlike the series type, the parallel structure does not process single sensor data but import all the sensors' data into the data fusion center for computation to obtain the final result. The hybrid fusion type is shown in Fig. 6.7c. The data from all the sensors may be input into different data fusion center. The data from the lower level data fusion center is then fused into the upper level data fusion center to obtain the final result.

## 6.2.3 Classification of Data Fusion

There are mainly two methods to classify the data fusion hierarchy: the first method refers to the classification based on the different level of target fusion which classifies the data fusion into low level (data level or pixel level), intermediate level (feature level) and high level (decision level); the second method refers to the classification of the sensor integration and data fusion into signal level, evidence level and dynamic level.

### 6.2.3.1 Data Level Fusion

The raw data of the sensor and the information generated at each stage of the pre-processing separately goes through fusion processing. By retaining most of the original information, this provides the minute information to the two other layers for fusion. At the same time, there exists limitations to data level fusion: (1) The sensor information that requires processing is large, and hence the high processing cost; (2) As the fusion is conducted at the lowest data level and the nature of the sensor's raw data is of uncertainty, incompleteness and instability, it requires a relatively high capacity for error correction during the fusion process; (3) sensor's data is required to be homogenous in quality as the information from each sensor needs to be precise at pixel level, and hence same quality type of sensors are required; (4) high transmission volume (Fig. 6.8).

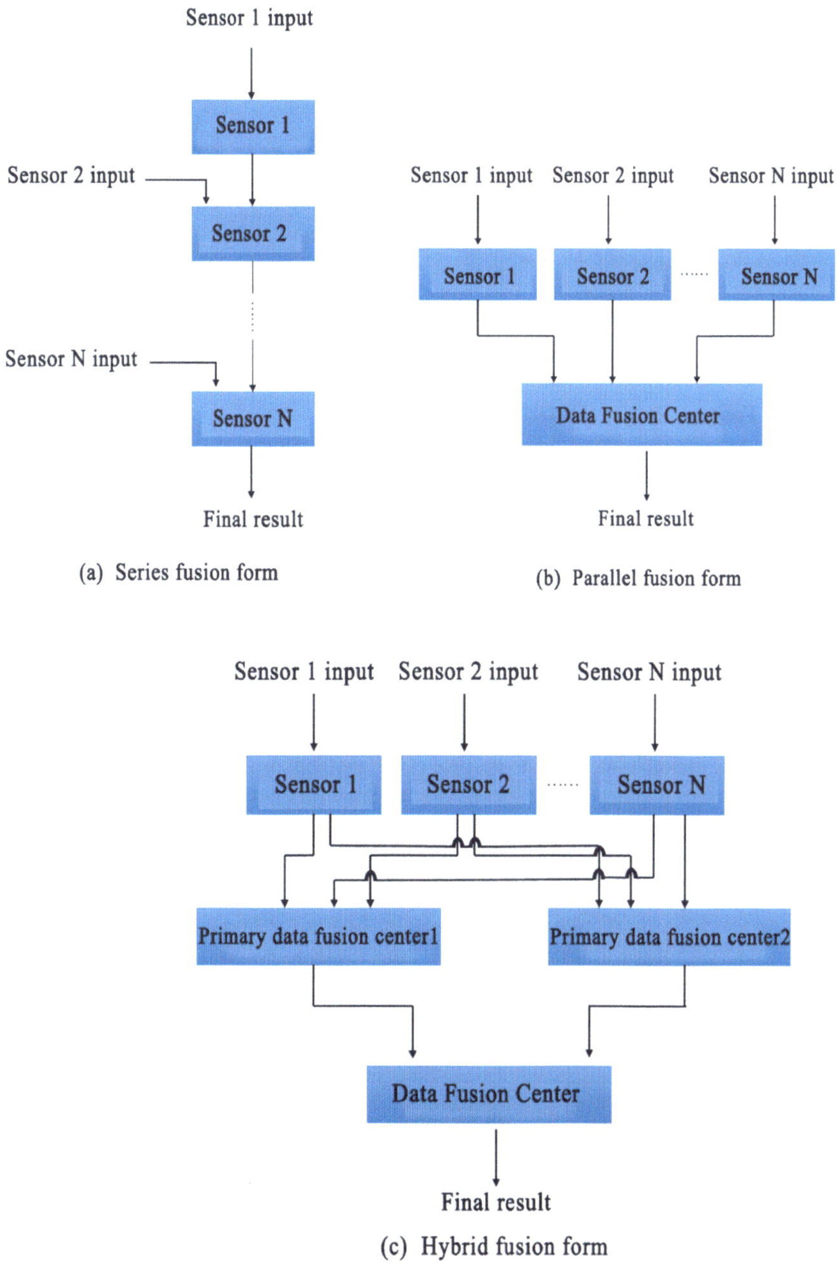


**Fig. 6.7** Data fusion architecture

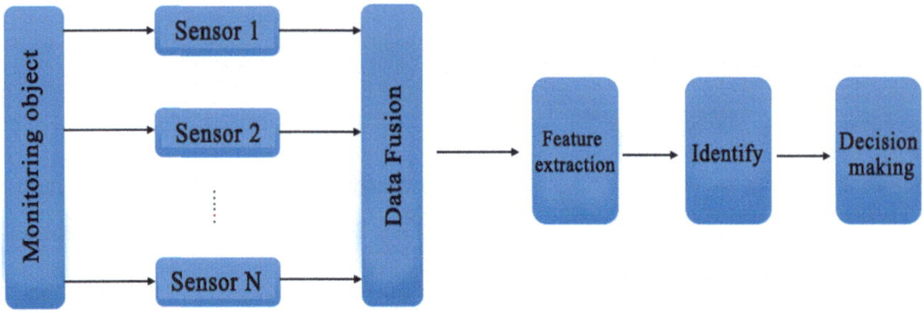


**Fig. 6.8** Process of data level fusion

#### 6.2.3.2 Feature Level Fusion

By using the feature information that is extracted from the respective raw sensor data, a comprehensive analysis and processing is conducted at the intermediate level. Generally, the selected feature information should be the comprehensive data representation or statistical quantity, therefore, it is used to perform classification, collation and integration on the sensor data. Feature level fusion is classified into target state information fusion and target feature fusion (Fig. 6.9).

1. **Target State Information Fusion**

Target state information fusion is mainly used in the field of multi-sensor target tracking. The fusion system first pre-processes the sensor data to complete the data registration. After data registration, the fusion processing mainly implements parameter correlation and state vector estimation.

2. **Target Feature Fusion**

In feature level fusion recognition, the specific fusion method adopted is the pattern recognition technology. This means the feature needs to be pre-processed before the fusion, that is, to classify the feature vector as appropriate. In the fields of pattern

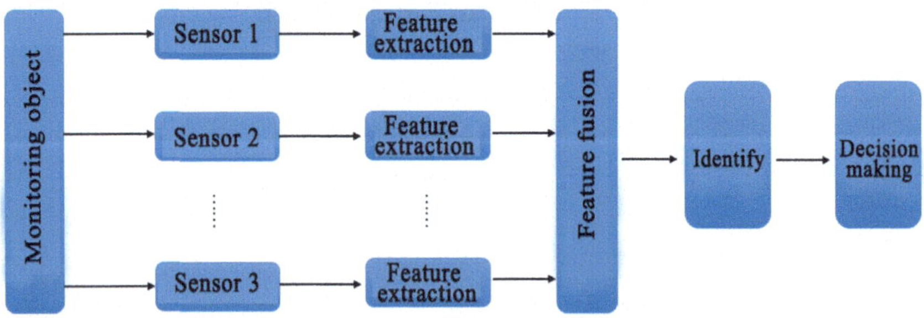


**Fig. 6.9** Process of feature level fusion

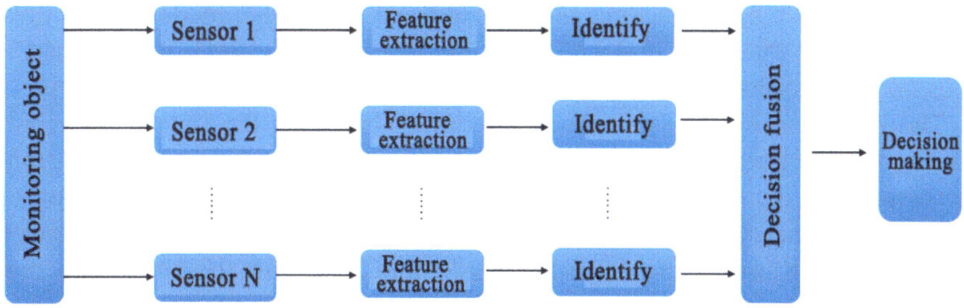


**Fig. 6.10** Decision level fusion

recognition, image processing and computer vision, there are already existing in depth studies regarding feature extraction and feature classification in which many methods are available for use.

#### 6.2.3.3 Decision-Level Fusion

This refers to the representation of data at the highest level of fusion processing. Different types of sensors will observe the same target. Each sensor performs pre-processing, feature extraction, recognition or judgment locally to establish preliminary conclusion on the observed target. Then, the joint inference results are obtained through correlation processing and decision-level fusion judgment. Thus, providing a basis for decision making.

Decision-level fusion directly targets specific decision-making objectives, making full use of the various types of feature information obtained by feature-level fusion to provide concise and intuitive results. Decision-level fusion gives the best real-time result and is also able to output a final decision even when one or several sensors fail. Hence, it has a good tolerance for errors (Fig. 6.10).

## 6.3 Multi-sensor Data Fusion

### *6.3.1 Kalman Filter*

#### 6.3.1.1 Concept

Due to the limitation of the sensor and the presence of noise in the natural environment, any measurement that we obtain from the sensor will consist of error, that is, we usually cannot accurately know the current state of the object. Most of the time when we wish to estimate the state of a particular object, we would conduct measurement. However, we are unable to fully trust our measurement as it is not accurate due to the presence of noise. At this stage, by using the sensor's measurement as a basis, tracking and estimation of the target's motion state would be performed to ensure the obstacle's position, speed and other relevant information does not change abruptly and is accurate.

The basic concept of the Kalman filtering is based on the state space representation of the linear system to find the optimal estimate of the system state from the output and input observation data. The system state refers to a collection of minimum parameters that summarize the effects of all the past input and disturbance on the system. Knowing the state of the system, together with the future input and system disturbance, can determine the overall behavior of the system.

Kalman filtering does not require the assumption of both signal and noise to be a stationary process. For system disturbances and observation errors (i.e., noise) at each moment, as long as some appropriate assumptions are made about their statistical properties, by processing the observation signals containing noise, the estimate of the true signal with the smallest error could be obtained on average.

For instance, in the unmanned driving system's environment sensing and target tracking application, the measurement of the target's positioning, speed and acceleration often exists noise at any time. The Kalman filtering makes use of the target's dynamic information to remove the impact of noise to obtain a good estimate of the target's positioning. This estimate could be the target's current positioning (filtering) estimate, the future positioning (prediction) estimate or the previous positioning (interpolation or smoothing) estimate.

The following section will discuss the Kalman filter in the engineering system.

In this section, the Kalman filter which originated from Dr. Kalman is being described. The following description involves some basic mathematical knowledge including probability, random variables, Gaussian or normal assignments, etc. However, the detailed proof derivation of the Kalman filter will not be covered here.

First and foremost, a system of discrete control processes is introduced. The system can be described by a linear stochastic differential equation:

$$X(K) = AX(k-1) + BU(k) + W(k) \tag{6.1}$$

Adding the system's measured value:

$$Z(k) = HX(k) + V(k) \tag{6.2}$$

In the above two equations, X(k) is the system's state at time; U(k) is the system's control quantity at time k; A and B are the system parameters which are matrices for multi-model systems; Z(k) is the measured value at time k; H is the parameter of the measurement system where H is a matrix in multi-measurement systems; and W(k) and V(k) represent the noise during the process and the measurement respectively. They are assumed to be the Gaussian white noise and their covariances are Q and R respectively (here we assume that they do not change with the system state changes).

In satisfying the above-mentioned conditions (in the linear stochastic differential systems, both the process and measurement consist of Gaussian white noise), the Kalman filter is the most optimal information processor. In the following, the covariances are combined to estimate the optimal output of the system (similar to the temperature example in the previous section).

Firstly, we need to use the system's process model to predict the next state of the system. Assuming that the current system state is k, depending on the model of the system, the current state can be predicted based on the previous state of the system:

$$X(k|k-1) = AX(K-1|K-1) + BU(k) \tag{6.3}$$

In Eq. (6.3), X(k|k − 1) is the predicted result based on the previous state; (k − 1|k − 1) is the optimal outcome of the previous state; U(k) is the controlled amount of the current state. If there is no control amount, U(k) will be equivalent to zero.

Until this stage, the system results have been updated, but the covariance corresponding to X(k|k − 1) has not been updated. We use P to represent the covariance:

$$P(k|k-1) = AP(k-1|k-1)A' + Q \tag{6.4}$$

In Eq. (6.4), P(k|k − 1) is X(k|k − 1) corresponding covariance; P(k − 1|k − 1) is X(k − 1|k − 1) corresponding covariance; A' represents the transpose matrix of A; and Q is the covariance of the process system. Equations (6.3) and (6.4) are namely two out of the five formulation of Kalman filter, that is, for system prediction.

Now that we have the predictions for the current state, we then collect the measurements for the current state. Combining the predicted and measured values, we can get the optimal estimate X(k|k) of the current state (k):

$$X(k|k) = X(k|k-1) + K(k)(Z(k) - HX(k|k-1)) \tag{6.5}$$

where K is the Kalman gain:

$$Kg(k) = \frac{P(k|k-1)H'}{HP(k|k-1)H' + R} \tag{6.6}$$

Up to now, we have obtained the optimal estimate X(k|k) in the k state. But in order to keep the Kalman filter running until the end of the system process, we also need to update the covariance of X(k|k) in the k state:

$$P(k|k) = (I - K(k)H)P(k|k-1) \tag{6.7}$$

wherein I is 1 matrix, and for a single model measurement, I = 1. When the system enters the k + 1 state, P (K|K) is P (1 − K|1 − K) in Eq. (6.4). In this way, the algorithm can continue in an autoregressive operation.

#### 6.3.1.2 Advantages of Kalman Filter in Data Fusion

As mentioned at the beginning of this chapter, an unmanned vehicle usually uses a variety of sensors: stereo cameras are often used to acquire image and distance information; traffic sign cameras are used for identification of traffic signs; radars are typically installed in the front bumper of vehicles. It is used to measure objects relative to the vehicle coordinate system. It can be used for positioning, ranging, speed measurement, etc. It is easy to be interfered by strong reflective objects. It is usually not used for the detection of stationary objects. Lidar is often installed on the roof and uses infrared. The laser beam is used to obtain the distance and position of the object. Although the spatial resolution is high, it is cumbersome and easily affected by the weather.

It can be seen that various sensors have their advantages and disadvantages. In actual driverless cars, we often combine the data of various sensors to sense the environment around our vehicles. This process of combining measurement data from various sensors is called sensor fusion. Kalman filtering also has many uses in driverless driving.

Firstly, in driverless driving, Kalman filter can be used for state estimation, that is, mainly for state estimation of pedestrians, bicycles, and other vehicles. The diagram of the basic principle and concept is shown in Fig. 6.11.

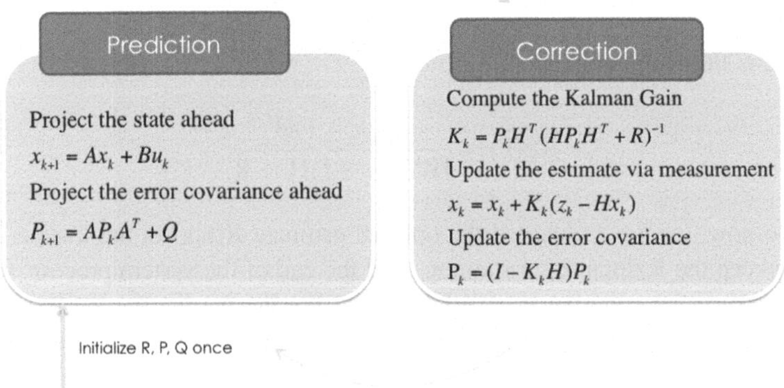


**Fig. 6.11** Concept of state estimation

Secondly, in unmanned driving, the Kalman filter can also combine the measured results from sensors such as radar and laser radar. The basic principle diagram is as follows (Fig. 6.12):

In addition, the Kalman filter is also of great use in the advanced motion model of vehicle tracking (Fig. 6.13).

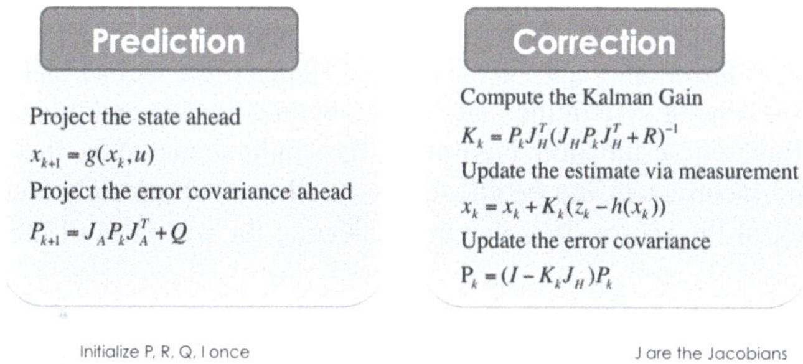


**Fig. 6.12** Concept of measurement

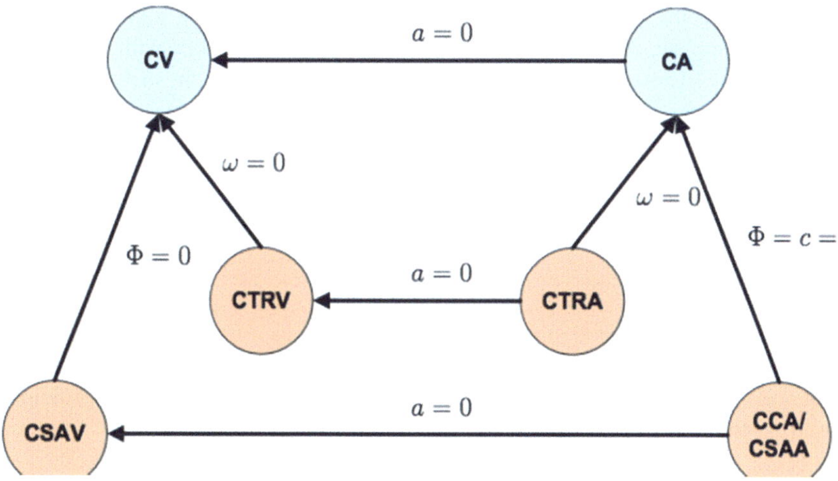


**Fig. 6.13** Advanced motion model

## 6.3.2 *Bayesian Estimation*

### 6.3.2.1 Description of the Bayesian Estimation Algorithm

(1) Fundamental Concept

In the unmanned driving system, as some portion of the original information in the signal is damaged due to noise or other factors and is unable to be recovered even after subsequent processing, as well as the fact that the signal itself is unable to carry the complete information of the observed target, and thus result in the incomplete and unprecise data provided by the multiple sensors, contributing to high amount of uncertainty and inconsistency. Hence these factors had resulted in the unmanned driving system the inability to obtain complete and precise data. By using the Bayesian estimation algorithm, the multi-source data that is highly uncertain and inconsistent can be effectively fused and substantially recognize the inconsistency of the sensor data, thereby, reflecting the true state of the measured data.

Bayesian estimation algorithm provides a means for data fusion. It is a common method for multi-sensor data fusion in a static environment, and has a strict theoretical basis. The information is described as a probability distribution, and is applicable for processing data with uncertainty that consists of the additive Gaussian noise (probability density function follows a Gaussian distribution).

In the multi-sensor data fusion system, as a part of the original information in the signal is damaged by noise and cannot be restored even in the subsequent data processing. In addition, as the signal itself is unable to contain the complete information of the observed target, this had resulted in many multi-sensors to provide incomplete, inaccurate, ambiguous and contradictory data with high levels of uncertainty. These factors had caused the data fusion system to be unable to adopt mathematical methodology for problem solving and to increase the content of the acquired data through signal post-processing. Hence, the data fusion centers have to further infer base on the uncertain information to achieve the objectives of target recognition and characteristics judgement. The Bayesian estimation algorithm can effectively fuse the multi-source uncertainty data.

(2) Fundamental Concept

The Bayesian estimation algorithm treats each sensor as a Bayesian estimator, and combines the associated probability distribution of each observed object into a joint posteriori probability distribution function according to the probability principle. Based on the observation values, the hypothesized likelihood function is continually being updated. The final fusion output of the multi-sensor data is then obtained by using the likelihood function of the joint distribution function. In this process, the likelihood function assumes a major role.

We know that the concept of likelihood is applied to estimate the parameters of a particular object of interest when certain observation values are known. In this sense, the likelihood function can be understood as the inverse of the conditional probability.

Knowing a certain parameter θ, the probability that an event will occur is given as P(A|θ), where:

$$P(A|\theta) = \frac{P(A,\theta)}{P(\theta)} \quad (6.8)$$

Using Bayes' Theorem,

$$P(\theta|A) = P(A,\theta)\frac{P(\theta)}{P(A)} \quad (6.9)$$

If the parameter estimation problem is discussed in the case of an unknown parameter, θ, as a non-random variable and some additional information of the unknown parameter can be provided in advance, it will be beneficial to estimate the parameter θ. By setting θ as the parameter of the overall distribution ρ(x|θ), in order to perform parameters estimation, a random sample, $X = (x_1, x_2, \ldots x_m)$, from the general distribution could be extracted. At the same time, based on θ's a priori information selected by a prior distribution, π(θ), by applying the Bayesian formulation to obtain the posterior distribution, π(θ|x), the estimation of θ could be conducted by using a location feature value from π(θ|x). Hence, the estimation is the simplest form of inference by using the posterior distribution. This is the basic concept of the Bayesian estimation.

(3) Fundamental Methodology

The Bayesian estimation method can fuse multi-sensor information to calculate the posterior probability that a given hypothesis is true. Setting n number of sensors, which can be of different types, and perform detection on the same target. The set target has m number of attributes that requires identification, that is, there are m number of hypotheses or propositions, $A_i$ where i = 1, … m (Fig. 6.14).

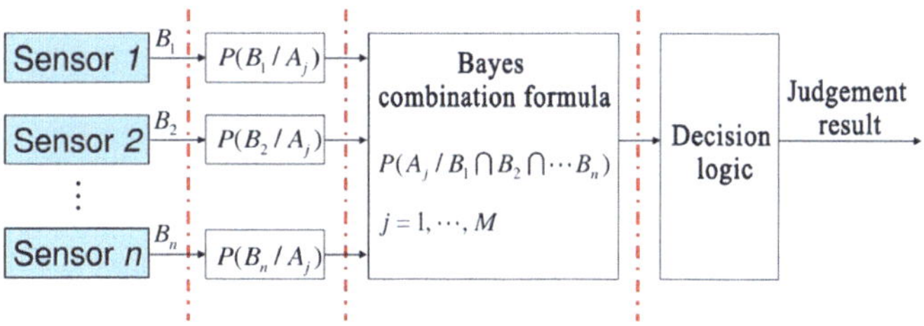


**Fig. 6.14** Process of fusion reasoning based on Bayesian estimation

#### 6.3.2.2 Limitations of the Estimation Algorithm

Although the Bayesian estimation algorithm manages to solve some of the shortcomings of the traditional estimation method, it also has several limitations, which are summarized into the following three aspect.

(1) Difficult to define a priori likelihood probability

The Bayesian estimation algorithm can directly determine the probability that the hypothesis is true given a parameter, and allows the hypothesis to be used to determine the likelihood prior knowledge, allowing the use of subjective probability as the a priori probability of the hypothesis. However, it is necessary to provide the distribution type and the a priori likelihood probability of the different types of sensor-observed objects in advance. The type of distribution is difficult to determine or it is not accurate enough. The a priori likelihood probability is based on a large amount of statistical data. When the processing problem is complicated, it requires a very large statistical workload, often requires the solving of a likelihood equation, wherein there are also problems such as insufficient precision, which makes it difficult to define the a priori likelihood probability. Thus, in many practical problems, the a priori likelihood probability generally relies on experience, with a larger focus on subjective factors.

(2) Unable to handle generalized uncertainty problem

The Bayesian estimation algorithm requires incompatible or independent hypothetical events. When there are multiple potential hypotheses and multi-condition related events, some assumptions are required to be mutually exclusive, which makes the computational complexity increase rapidly, reflecting the lack of Bayesian estimation algorithm's ability to allocate the total uncertainty.

(3) Cannot directly use the Bayesian estimation algorithm for data fusion

When the observation coordinates of the sensor group are the same, the data of the sensor can be directly fused, but in most cases, the sensor describes the same object from different coordinate frames. At this time, the observation data requires the indirect use of the Bayesian estimation for data fusion. The problem to be solved by the indirect method means to find the rotation matrix and translation vector that is consistent with the group of sensors.

### *6.3.3 D-S Evidence Theory*

D-S evidence theory was first proposed by Dempster in the twentieth century in the 1960s. Subsequently, Shafer took on to further expand and develop the theory. The evidence theory focusses on the outcome of an event to seek the main causes of the event occurrence. Evidence theory is a method for uncertainty reasoning. It allows a problem and evidence to be broken down into several sub-problems and sub-evidences. After the subsequent processing of the sub-problems and sub-evidences, the synthetic method is adopted to obtain the solution to the original problem. At the same time, evidence theory has a strong theoretical basis. The description of uncertainty in decision-making is more in line with people's habitual thinking, and it can reduce the set of hypotheses and integrate the opinions of multiple decision makers through the accumulation of evidence and the rules of synthesis. Therefore, it is often used as a decision-making method in multi-attribute group decision making, and it is also widely used in unmanned autonomous mobile systems.

#### 6.3.3.1 Basic Concept

Evidence theory describes uncertainty by identifying concepts such as framework, basic probability distribution function, belief function, likelihood function, and correlation. In evidence theory, the discrete value range of the research object is called the discernment framework. The mutually exclusive and exhaustive elements in the discernment framework are called primitives.

Set $U$ as the frame of discernment, if the function $m : 2^U \to [0, 1]$ ($2^U$ is the set of all subsets of $U$) satisfies the conditions below:

(1) $m(\phi) = 0$, $\phi$ as the empty set or the set of impossible events;
(2) $\sum_{A \subseteq U} m(A) = 1$;

And $m(A)$ is the basic probability distribution function of $A$. $m(A)$ represents the belief precision degree of $A$, indicating the direct support for $A$.

Define the function $BLE$: $2^U \to [0, 1]$ as:

$$BLE(A) = \sum_{B \subseteq A} m(B), \quad \forall A \subseteq U \tag{6.10}$$

And $BLE(A)$ is known as the belief function of $A$, which represents all the degrees of belief for the hypothesis $A$ and all the subsets of $A$. Numerically, it is equivalent to the basic credibility of the hypothesis $A$ and all the subsets of $A$.

If a subset A in a frame of discernment $U$ has $m(A) > 0$, then $A$ is a focal element of the belief function $BLE$. All focal elements are known as cores.

It is insufficient to describe the belief of a proposition $A$ only by the belief function because $BLE(A)$ does not reflect the degree of rejection of the hypothesis. This means that the belief level of $A$ is not true. Hence, in order to be able to describe the belief of $A$ holistically, it is necessary to introduce a likelihood function to indicate the level of skepticism towards $A$. Its definition is as follows:

Set $U$ as a frame of discernment, set the function $PL$: $2^U \to [0, 1]$ as:

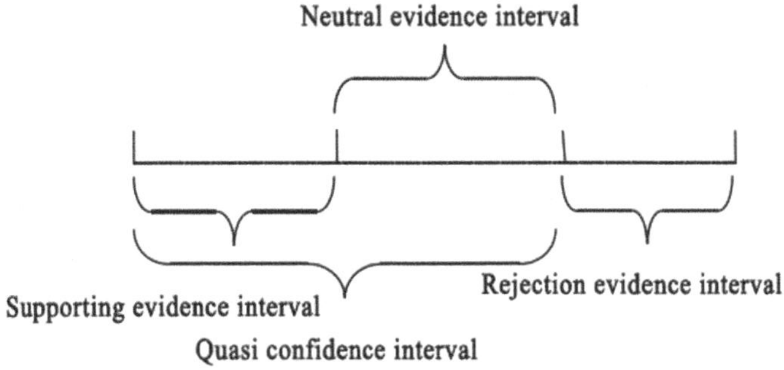


**Fig. 6.15** Three types of evidence

$$PL(A) = \sum_{B \cap A \neq \phi} m(B) \tag{6.11}$$

Then, $PL(A)$ is known as the likelihood function. This expresses the non-false level of confidence of A, or also known as the level of non-suspicion of $A$.

As shown in Fig. 6.15, pertaining to the evidence of the proposition $A$, there are three types of evidence: supporting evidence, rejection evidence and neutral evidence. Of which, supporting evidence refers to evidence that supports the proposition $A$; rejection evidence refers to evidence that rejects the proposition $A$; and neutral evidence refers to evidence that could neither directly support nor reject the proposition $A$. As both the supporting and neutral evidence could possibly support the proposition $A$, hence, the evidence theory refers to both types of evidence as the fiduciary interval.

#### 6.3.3.2 Combination Rule of Evidence Theory

Based on the above-mentioned concepts, the belief function and the likelihood function could be used to measure the uncertainty of the proposition. The definition of the belief function and the likelihood function both depend on the basic belief distribution function, as such, the basic belief distribution function becomes the foundation of the uncertainty of a proposition. Nonetheless, although the same evidence is regarded, a different basic belief distribution function is obtained as the data source is different. In such a situation, in order to compute the belief function and the likelihood function, the different basic belief distribution functions must be combined into a probability distribution function. Hence, Dempster and Shafer proposed a synthesis method to perform orthogonal sum operation on two or more basic belief distribution functions. The formulation of the synthesis method is known as the Dempster-Shafer synthesis rule, or in short, the D-S synthesis rule.

Set $BLE_1$ and $BLE_2$ as two belief functions in the same frame of discernment $U$. The basic probability distribution functions are namely $m_1$ and $m_2$, while $A_1, \cdots A_K$ and $B_1, \cdots B_r$ are the corresponding focal elements respectively. Thus,

$$k = \sum_{\substack{i,j \\ A_i \cap B_j = \phi}} m_1(A_i) m_2(B_j) < 1 \tag{6.12}$$

The Dempster rule that synthesize the two sets of evidence is as follow:

$$m(C) = \begin{cases} \frac{\sum_{\substack{i,j \\ A_i \cap B_j = \phi}} m_1(A_i) m_2(B_j)}{1-k} & \forall C \subset U, C \neq \phi \\ 0 & C = \phi \end{cases} \tag{6.13}$$

In the formula, $k$ is the degree of conflict between evidences, known as the conflict coefficient, which indicates the amount of evidence conflict that is measured conventionally. The larger the $k$ value,the larger is the evidence conflict. When the $k$ value approaches towards 1, the use of the Dempster-Shafer rule to fuse high-conflict evidence will lead to a contradictory conclusion. The above operations using the Dempster-Shafer rule are also called quadrature and (or straight sum), and is written as:

$$m = m_1 \oplus m_2 \tag{6.14}$$

#### 6.3.3.3 Decision Rules for D-S Evidence Theory

Upon completing evidence synthesis, how to perform decision making is another important problem. The choice of decision-making methods often needs to be set separately according to different application environments. Under the discernment framework where $m$ is the basic evidence probability assignment after synthesis, the general decision-making methods include: decision based on probability assignment, decision based on belief function and decision based on minimum risk. The following is a brief introduction to decision-making based on probability assignment.

Set $\exists A_1, A_2 \subset U$, to satisfy:

$$m(A_1) = \max\{m(A_i), A_i \subset U\} \tag{6.15}$$

$$m(A_2) = \max\{m(A_i), A_i \subset U \textit{ and } A_i \neq A_1\} \tag{6.16}$$

If there are

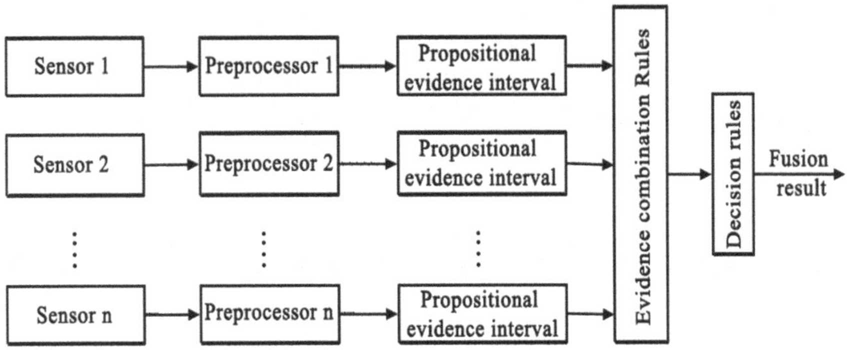


**Fig. 6.16** Fusion model based on evidence theory

$$\begin{cases} m(A_1) - m(A_2) > \varepsilon_1 \\ m(U) < \varepsilon_2 \\ m(A_1) > m(U) \end{cases} \tag{6.17}$$

Then, $A_1$ is the result after judgement,where $\varepsilon_1, \varepsilon_2$ are preset thresholds.

#### 6.3.3.4 Data Fusion Model Based on D-S Evidence Theory

The unmanned autonomous mobile system is generally equipped with a variety of sensors. According to the DS evidence theory data fusion algorithm, the input values of multiple sensors can be processed to obtain a fusion result. The basic process is shown in Fig. 6.16.

1. Pre-process the detected data from each sensor;
2. Calculate the basic probability assignment, credibility and fiduciary of each data;
3. Based on the derivation of the Dempster synthesis rule, pertaining to the fusion of all the evidence information for the same problem, obtain the synthesized basic probability assignment, credibility and fiduciary; and
4. Derive a conclusion based on the D-S evidence theory decision rule.

### *6.3.4 Fuzzy Logic Inference*

#### 6.3.4.1 Basic Concept and Theory of Fuzzy Logic

It is common for the unmanned systems to utilize multiple sensors to perform environmental awareness. Under different time periods and working environment, the reliability of the output varies with the different sensors. In the practical application of the health monitoring of the unmanned system, measurements from a single sensor is unlikely to be correct. Under normal circumstances, due to the inaccuracy of a single sensor and the presence of other factors such as the environmental noise and human interference, the fused health monitoring data often has a certain degree of uncertainty. It is mainly manifested by the incompleteness of the health monitoring data, the unreliability of obtaining the health monitoring data, the immaturity of representing the health monitoring data and the contradiction among the various health monitoring data.

A sensor may produce false or even erroneous measurement data due to various reasons in a measurement. Such measurement data is known as the wild value. If the wild value is sent to the health monitoring data fusion center, the accuracy of the fusion outcome will be affected. Hence, prior to performing the health monitoring data fusion, the measured values from all the sensors must be judged and identified so as to search and eliminate the wild values to ensure the validity of the data from each sensor and the accuracy of the fusion result.

In order to solve the uncertainty problem due to data from the various sensors during the multi-sensor data fusion process in the unmanned system, the fuzzy information processing method is usually adopted. Fuzzy information processing can effectively utilize the effective information of all sensors to ensure that the effective information is not lost and the influence from information with large error is reduced, so as to ensure the information fusion precision of the whole system. The correctness of the measured data from each subsystem is detected by the probability distribution theory and the adaptive variable weight analysis method. Thereby, a new multi-sensor fusion data validity fuzzy evaluation method is proposed. Specifically, it utilizes the multiple measurements of the various sensors on a subsystem of the unmanned health monitoring system for fuzzy identification to obtain a comprehensive evaluation. This method enables the selection and the adaptive adjustment by the membership function, and to judge the validity of the output data of each subsystem more accurately.

The basic idea of the fuzzy fusion is to obscure the absolute membership in the classical set. From the aspect of feature function, the degree of membership of element x is no longer limited to only 0 or 1 choices, but can take any value between 0 and 1.

Fuzzy set membership: the fuzzy set A in the universe X is characterized by the membership function $\mu A(x)$. $\mu A(x)$ takes the values on the real axis of the close interval [0, 1]. The value of $\mu A(x)$ represents the membership degree that the element x in universe X belongs to the fuzzy set A. The real number, fuzzy set R of the membership function is usually known as the fuzzy distribution. The concept of fuzzy set eliminates the concept of "one or the other" in the ordinary set, and effectively plans the uncertainty of the physical classification boundary in the real world. When the objective fuzzy phenomenon in question is similar to a given fuzzy distribution, this fuzzy distribution can be selected as the membership function sought. Figures 6.17, 6.18, and 6.19 shows several common fuzzy distributions and their graphs.

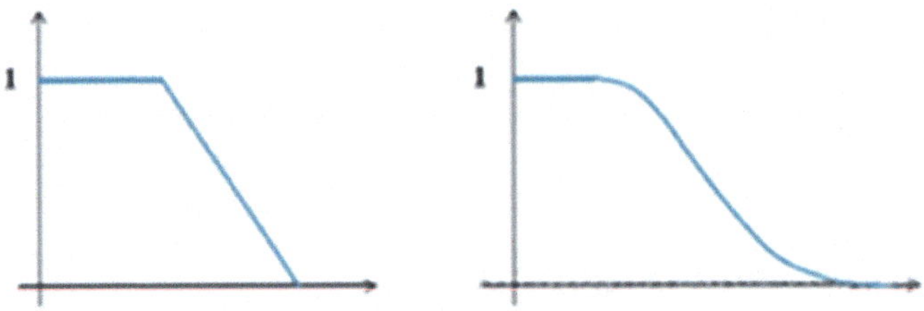


**Fig. 6.17** Z-shaped (generally indicate fuzzy sets on the leftmost coordinate system)

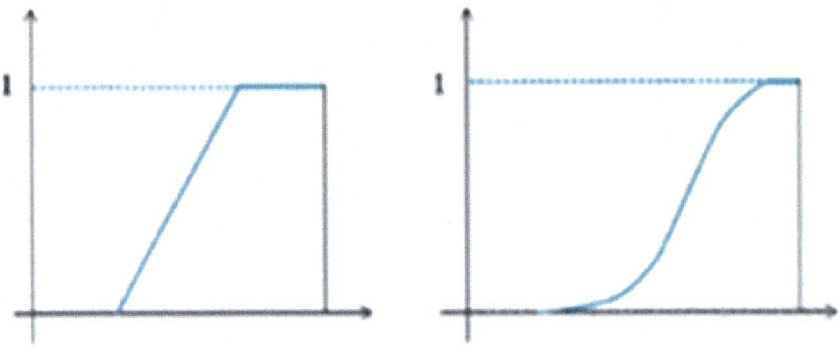


**Fig. 6.18** S-shaped (generally indicate fuzzy sets on the rightmost coordinate system)

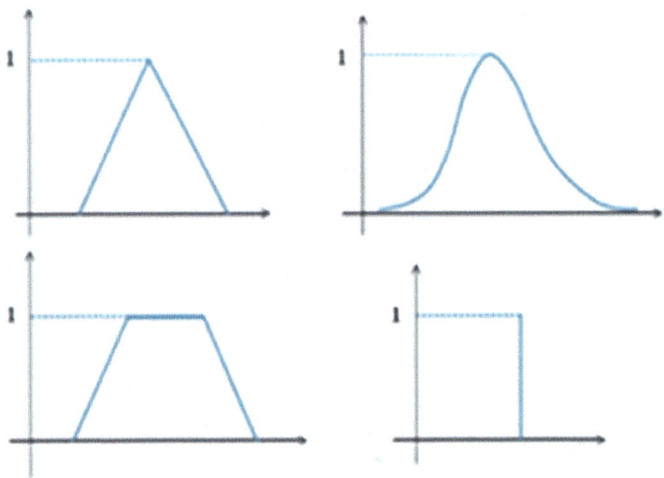


**Fig. 6.19** Convex shaped (generally indicate fuzzy sets in the middle region of the coordinate system)

Fuzzy logic is a kind of multi-value logic, and the degree of membership can be regarded as an inexact representation of the true value of a data. Hence, the inaccuracy that exists during the multi-sensor data fusion process could directly be represented by the fuzzy logic. Thereafter, the multi-value fuzzy logic could be applied for reasoning and then perform fusion on the data from multiple sensors or multiple data from a single sensor for a comprehensive evaluation of the result. It can be seen that this fuzzy information processing technology developed by fuzzy set theory can provide a simple and effective means for the exploration of uncertainty problems and the simulation of human reasoning. Since fuzzy systems can structure human empirical knowledge, they can generally provide a clear explanation of their operational behavior, so that each parameter has a clear physical meaning.

The fuzziness of a fuzzy subset is usually described by the magnitude of the ambiguity. Ambiguity is the overall description of a fuzzy element in a fuzzy set of a particular universe which quantitatively analyze the degree of fuzziness of the fuzzy set.

Ambiguity: The ambiguity $D(A) \in [0, 1]$ of the fuzzy set A in the Universe X refers to:

(1) For $\forall x \in X$, when $A(x) = 0$ or 1, A degenerates into a classic set, then $D(A) = 0$ and the ambiguity becomes the minimum;
(2) When $A(x) = 0.5$, $D(A) = 1$, the ambiguity becomes the maximum;

Pertaining to the two fuzzy sets $A_1(x)$ and $A_2(x)$ in X, if $A_1(x) \geq A_2(x) \geq 0.5$ or $A_1(x) \leq A_2(x) \leq 0.5$, then $D_1(x) \leq D_2(x)$;

(3) For any $A \in F(X)$, $D(A) = D(\overline{A})$ is established.

#### 6.3.4.2 Multi-sensor Data Fusion Models Based on Fuzzy Evidence Theory

In the problem of multi-sensor data fusion, how to reasonably determine the weight coefficient of each sensor in the fusion process based on the measured values to ensure the accuracy of the fusion result is the focus of research. Based on the concept of the evidence theory where the basic belief assignment is taken as reference and combined with the fuzzy logic, the reliability of the measurement from each sensor is being evaluated via the degree of membership and mutual support between the measured values which is then converted into evidence. Thereafter, the D-S evidence theory synthesis rule combines all the evidences to obtain a weight distribution function for each measured value. The multi-sensor data fusion model based on the fuzzy evidence theory is shown in Fig. 6.20.

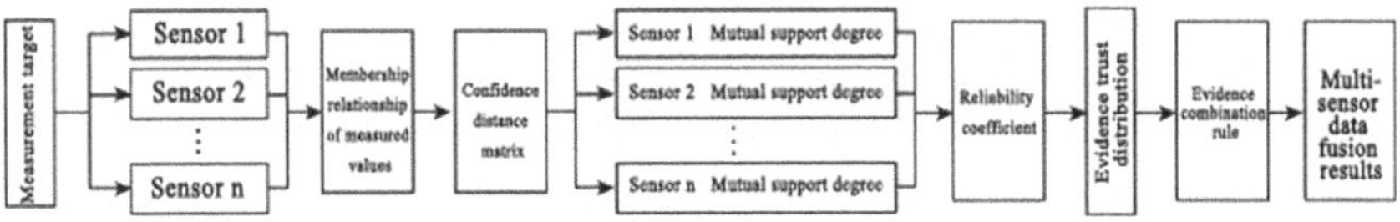


**Fig. 6.20** Multi-sensor data fusion model based on fuzzy evidence theory

#### 6.3.4.3 Validity Evaluation of the Multi-sensor Data Fusion

The fuzzy processing of the operating state parameters in an unmanned system refers to the quantification of the measured data from each of the various sensors in the subsystem. Based on the influence of the operating state parameters of the unmanned system on the accuracy of the information fusion, the membership function and the correlation analysis method can be used to perform the fuzzy processing of the measured values of the various sensors in the subsystem. At present, the methods to determine the membership function are based on experience. The process includes the continuous correction of the understanding through feedback during the implementation so as to achieve the intended target. In many cases, it is easiest to approximate some fuzzy variables by using some common distribution functions as membership functions.

The effective fuzzy evaluation of the multi-sensor fusion data of the unmanned health monitoring system is usually based on the fuzzy value of each sensor in the subsystem to obtain the local decision of each sensor, and then to give a variable weight. The fuzzy operator calculates the reasonable judgment of each subsystem data, and finally uses the corresponding rules to carry out the final information fusion according to the judgment results of each subsystem.

### *6.3.5 Artificial Neural Network*

#### 6.3.5.1 Basic Concept

Artificial Neural Network (ANN), also known as neural networks, is designed based on the research of the biological neural networks of humans. Simple models are abstracted according to the process of how the human brain processes information. Different networks are formed based on different connections.

After nearly half a century of development, artificial neural networks rely on their self-learning capabilities, associative storage functions and the ability to find optimized solutions at high speeds to achieve extensive success in many research fields such as pattern recognition, automatic control, signal processing, unmanned systems, assisted decision making and artificial intelligence.

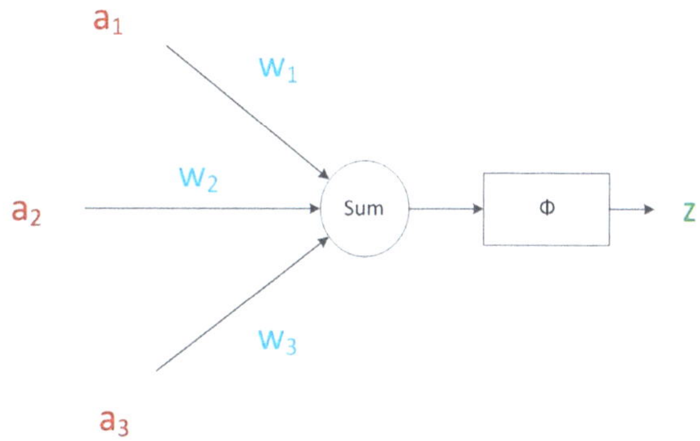


**Fig. 6.21** Artificial neuron

Based on the deep understanding of the working principle of the biological neural networks, people have abstracted the famous McCulloch—Pitts (MP) model, which is the simplest artificial neural network model.

In Fig. 6.21, $a_1 \sim a_3$ represent the input which refer to the external signal that is commonly received by the various sensors in the unmanned system data fusion model; $w_1 \sim w_3$ represent the weight of each synapse of the artificial neurons. The input signals are weighted and summed, that is, the cumulative effect of the input signals. Since each biological neuron has a specific threshold, the cell body will emit an electrochemical pulse only when the input signal is weighted and summed to exceed the threshold. Therefore, an activation function $\Phi$ is introduced into the artificial neurons to process the input signals obtained by the neurons, and eventually to obtain the output value z. The common activation functions include the linear functions, nonlinear bevel functions, step functions, sigmoid functions, etc. After the sensor signal is input into the artificial neural network model and the function of the weighted summation and the activation function is performed, the information fusion for the multi-sensor signal is completed, which will finally output the decision.

Abstracting the above diagram into a formula:

$$z = \Phi(a_1 \times w_1 + a_2 \times w_2 + a_3 \times w_3) \tag{6.18}$$

In general, an artificial neural network model has multiple inputs and outputs, so the above formula can be written as a vector:

$$\mathbf{z} = \Phi(\mathbf{w} \cdot \mathbf{a}) \tag{6.19}$$

Of which, **a** is the input column vector of n columns, **z** is the output column vector of m columns, and **w** is a coefficient matrix of m rows and n columns.

Since the M-P model only contains the input layer and the output layer, it can only solve the simple linear classification task. Therefore, it is often necessary to introduce one or more computing layers between the input layer and the output layer to form a two-layer neural network or even a multi-layer neural network.

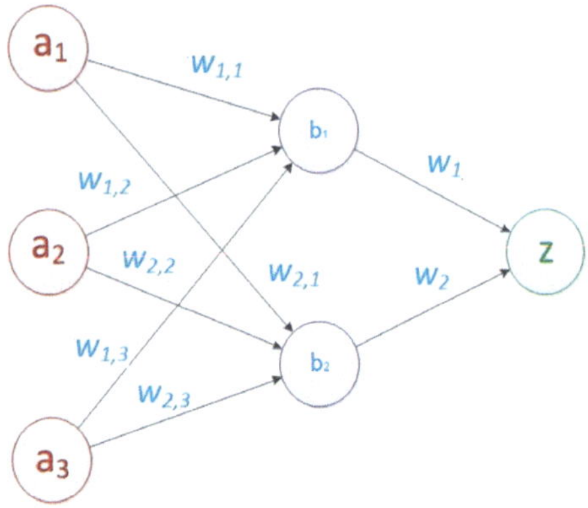


**Fig. 6.22** Two-layer neural network

The calculation layer introduced between the input layer and the output layer is called a hidden layer, and the number of hidden layer nodes does not depend on the number of nodes of the input and output layers, but is determined by the designer, and is generally set by the grid search method (Fig. 6.22).

The above diagram can be written as:

$$\mathbf{b} = \Phi(\mathbf{w} \cdot \mathbf{a}) \tag{6.20}$$

$$\mathbf{z} = \Phi(\mathbf{w}' \cdot \mathbf{b}) \tag{6.21}$$

In order to make the data fusion process more accurate given the input vector, it is necessary to determine and optimize the weight vector w in the neural network. This process is also called neural network training. In order to make the model as accurate as possible, it is necessary to make the difference between the actual output of the model and the theoretical output as small as possible.

Definition:

$$\mathrm{k} = (\mathrm{z_p} - \mathrm{z_o})^2 \tag{6.22}$$

where, $z_p$ is the theoretical output value of the sample and $z_o$ is the actual output value of the sample after passing through the neural network model. This function is also called the loss function. In order to make the loss function as small as possible, the general derivation calculation method will lead to a large amount of computation. Therefore, the back-propagation algorithm (BP algorithm) is often used. The basic idea is to use the gradient descent method and the gradient search technique. The loss function of this model is minimal, and such a two-layer neural network is also called a BP network (Fig. 6.23).

In a two-layer neural network, the activation function often uses the sigmoid function, and its expression is:

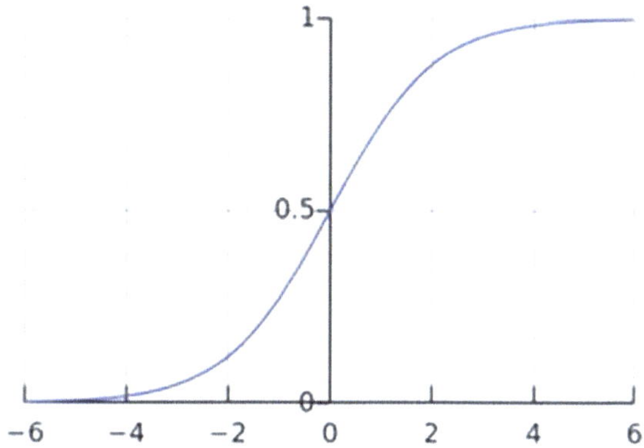


**Fig. 6.23** Sigmoid function

$$z = \frac{1}{1 + e^{-\text{sum}}} \tag{6.23}$$

where, z is the output value and sum is the weighted sum of the input signals. The function has a function value of 0.5 at sum = 0, and the value of the function changes greatly when sum is between (−0.6, 0.6), while the value changes little when the value is outside (−1, 1) (Fig. 6.24).

A multi-layer neural network is formed when one or more computing layers is (are) added in the two-layer neural network. Hence the formula is as follow:

$$\mathbf{b} = \Phi(\mathbf{w} \cdot \mathbf{a}) \tag{6.24}$$

$$\mathbf{c} = \Phi(\mathbf{w}' \cdot \mathbf{b}) \tag{6.25}$$

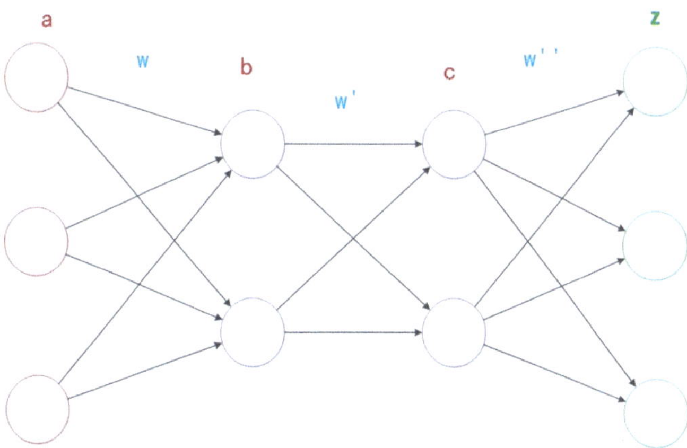


**Fig. 6.24** Multilayer neural network

**Fig. 6.25** ReLU function

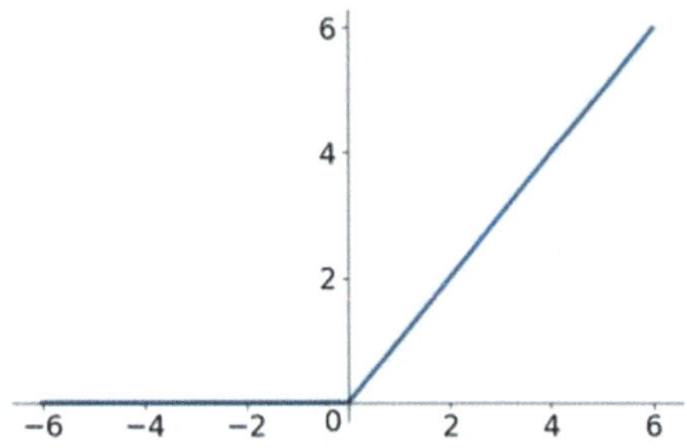


$$\mathbf{z} = \Phi(\mathbf{w}'' \cdot \mathbf{c}) \tag{6.26}$$

By increasing the number of layers in the neural network to obtain a deeper network, stronger representation capabilities and functional simulation capabilities can be obtained. In unmanned systems, the introduction of more computational layers enables deeper integration of sensor signals, closer output values to actual values, and more accurate decisions. In the training of the multi-layer neural networks, the gradient descent algorithm and the back propagation algorithm applied to the two-layer neural network can also achieve good results.

For the activation function, unlike the two-layer neural network, the multi-layer neural network ReLU function, is expressed as:

$$\mathrm{z} = \max(\mathrm{sum}, 0) \tag{6.27}$$

That is, when *sum* is greater than 0, the output value is the input *sum*, else the output is 0 (Fig. 6.25).

#### 6.3.5.2 Advantages of Artificial Neural Networks in Data Fusion

Neural networks are highly fault tolerant, self-learning, self-organizing, adaptive and capable of simulating complex nonlinear mapping. These characteristics of the neural network and the strong nonlinear processing capability nicely meet the requirements of the multi-sensor data fusion process. In multi-sensor systems, the environmental information provided by each information source has a certain degree of uncertainty, and the fusion process of these uncertain information is actually an uncertainty reasoning process. The neural network determines the classification criteria according to the sample similarity that is accepted by the current system. This determination method is mainly expressed in the weight distribution of the network. At the same time, the neural network specific learning algorithm can be used to acquire knowledge and obtain the uncertainty reasoning mechanism. By using the neural network signal processing capabilities and the automated reasoning function, a multi-sensor data fusion is realized.

# 6.4 Application of Multi-sensor Data Fusion

## 6.4.1 Application in Unmanned Aerial Vehicle

A variety of aircraft navigation devices has been created from modern science and technology, enabling the navigation system to evolve from a single sensor type to a multi-sensor integrated navigation system to achieve complementary performance. In the aspect of data processing, information processing of a data set from a single specific sensor is also developing towards the data fusion of multiple data sets from multi-sensors.

The data fusion of the unmanned aerial vehicle (UAV) navigation includes the fusion of the location and attribute, which is the lowest level of data fusion. The method is to describe the noise statistics of the system model and related sensors by state equation and observation equation, and then map the data to the state space for the state estimate. The state vector mainly includes position, velocity, angular velocity, attitude, etc. The flight state of the aircraft is determined by these parameters (Fig. 6.26).

In the process of autonomous take-off and landing of the UAVs, accurate height measurement is particularly important. This is one of the key technologies to realize the autonomous take-off and landing of the UAV. Since each sensor has its own advantages and disadvantages, a multi-sensor system is often used to measure a certain physical quantity in actual navigation. In UAV height measurement

**Fig. 6.26** Unmanned aerial vehicle

systems, height sensors include the GPS, sonar and barometric altimeters. GPS can provide global positioning information in real-time without interruption. However, in a high-speed dynamic environment, signal loss often occurs due to the limitation of the positioning principle. On the other hand, its data update frequency is low, and sometimes it is difficult to meet the requirements of navigation control. The deadliest is the possibility that the system positioning becomes invalid due to the US government's manipulation of the GPS system satellite. The main advantage of the electronic sonar is that it has extremely high precision in short-distance measurement. However, the measurement range is small, generally only within 10 m, and only the relative height can be measured. It is often necessary to know the local altitude while using it. Pneumatic height count data is continuous and has high resolution, and therefore it has a wide range of measurement. However, due to the change in atmospheric pressure, the height measurement accuracy will be reduced. There will be a certain measurement value error in different environments, which is subjected to environmental factors. The multi-sensor data fusion technology is applied to the UAV height positioning system to obtain higher-precision height values, which can realize the autonomous take-off and trajectory navigation of the UAV.

### *6.4.2 Application in Unmanned Vehicle*

Automated driving functions such as automatic parking, road cruise control and automatic emergency braking are largely dependent on sensors. It is not just the number or type of sensors that are important, but rather how they are used. At present, most of the ADAS in the vehicles on the road are working independently, which means that they exchange little information with each other. Combining the multiple sensor information is the key to achieving automatic driving.

The distinguishing feature of smart cars lies in artificial intelligence. This means that the car itself can sense the road environment through the on-board sensing system which automatically plans the driving route and controls the vehicle to reach the intended destination. At present, the vehicle sensing module includes a visual perception module, a millimeter-wave radar, an ultrasonic radar, a 360° surround vision system, etc. The multi-source sensor synergistically recognizes obstacles such as road lane lines and pedestrian vehicles for safe driving. Therefore, the perceived information also needs to be fused and complemented.

A variety of different sensors correspond to different working conditions and sensing targets. For example, the millimeter-wave radar is used to identify the obstacles such as road vehicles, pedestrians, road blocks, etc. in the distance range of 0.5–150 m. The ultrasonic radar identifies obstacles within a close proximity range of 0.2–5 m along the road during the parking process, in the front and rear of the stationary vehicle and pedestrians passing by, etc. The two work synergistically to complement the shortfall. By fusing the data which includes the obstacle angle,

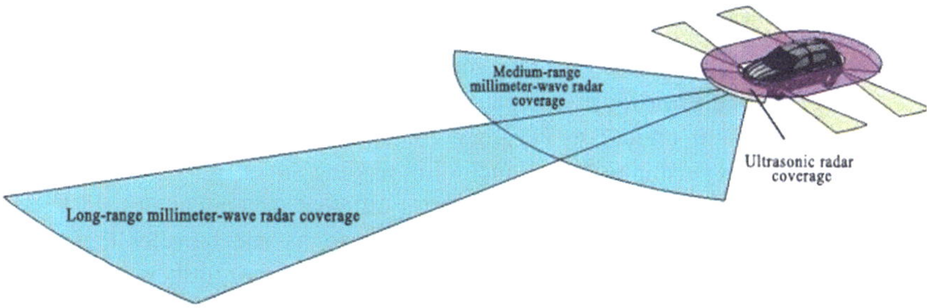


**Fig. 6.27** Smart car sensing module

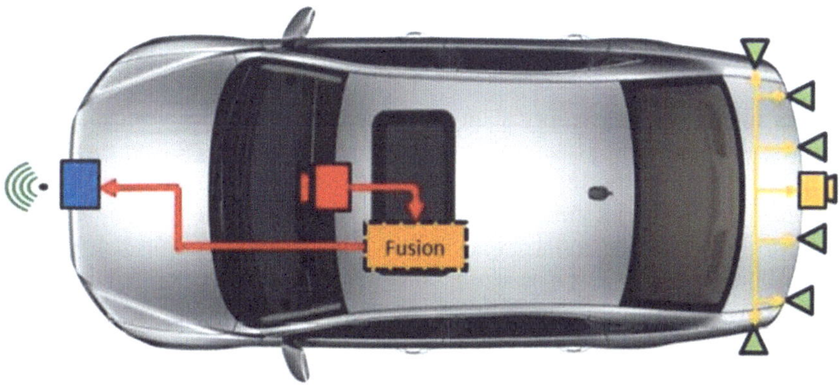


**Fig. 6.28** Example of unmanned vehicle sensor fusion

distance and speed, the surrounding environment of the vehicle body and the reachable space range are described (Fig. 6.27).

Two basic sensor fusion examples are: (1) rear-view camera and ultrasonic ranging; (2) front camera and multi-mode front radar (Fig. 6.28):

1. **Rear View Camera and Ultrasonic Ranging**

Ultrasonic parking assist technology is widely accepted in the automotive market and is very mature. A more advanced parking aid function can be achieved by combining a rear-view camera with the ultrasonic ranging. The rear-view camera allows the driver to clearly see the rear of the vehicle, while the machine vision algorithm can detect objects, as well as, shoulder stones and markings on the street. With the complementary function provided by the ultrasound, the distance of the recognized object can be accurately determined. In the case of low light or complete darkness, a basic proximity alarm can be ensured.

2. **Front Camera and Multi-Mode Front Radar**

Another powerful combination is the combination of the front camera function and the front radar. The front radar is capable of measuring the speed and distance of objects up to 150 m under any weather condition. The camera is excellent at detecting and distinguishing objects, including reading street signs and road signs. By using a multi-camera sensor that constitutes different viewing angles (FoV) and optical elements, the system may recognize pedestrians and bicycles that are in front of the vehicle, as well as, objects within 150 m range or further. At the same time, it can also reliably automate the emergency braking, the start and stop cruise control and other functions.

### 6.4.3 Application in Unmanned Surface Vehicle

In order to improve the sensory ability of the unmanned surface vehicle, the necessary sensors must be assembled according to the research requirements in the system design. However, relying on a single sensor cannot guarantee the accuracy and reliability of the information. Information fusion technology is the essential means to maximize the surface vehicle's perception ability. Compared with the information provided by a single sensor, after information fusion technology, it will be more accurate and reliable. Thus, accurately and quickly acquire the characteristics of the required information which provides the most reliable information source for a series of intelligent functions such as autonomous navigation and autonomous obstacle avoidance for the unmanned surface vehicle control platform.

In terms of ship navigation and obstacle avoidance, radar and the Automatic Identification the System (AIS) are more commonly used as the monitoring equipment. As each has its own characteristics, and hence they are unable to replace each other.

The advantage of the radar is that it has far observation distance and is not affected by factors such as foggy weather, which can influence the target position and direction quickly. The main disadvantage is that in the process of identifying and tracking the target, it is greatly affected by the clutter, the presence of blind zone, and low precision.

AIS has the advantages of high precision, strong anti-clutter and rich information content. The disadvantages are that only ships or shores that are also equipped with AIS equipment can be detected, is sensitive to radio frequency interference, is a passive tracking equipment, etc. (Fig. 6.29).

The track correlation can be realized by statistics. After the time calibration of the AIS and the radar equipment, the sampling time is consistent, and there is no possibility of correlation between the different time tracks. If there exists a correlation between the AIS target track and the radar target track, this means that the

**Fig. 6.29** Unmanned surface vehicle

distance between the two tracks would not be far from each other. Threshold filtering based on time and distance during track correlation will reduce the probability of error correlation. In this way, track correlation can be achieved.

## Bibliography


1. Li J, Jia L (2007) A survey of data fusion. Transp Res 9:192–195
2. Cheng Y, Huang X, Min (2005) Multi-sensor data fusion technique and its application. Sens Microsyst 24(10):1–4
3. Fan H-D, Lee SM (2002) Kalman filter algorithm geometric interpretation. Fire Control Command Control 27(4):48–50
4. Haykin SS (2001) Kalman filtering and neural networks. John Wiley & Sons, Inc
5. Cai Z, Zhang X, Niu C, Wei H (2017) A simplified strong tracking volume kalman filter algorithm. Electr Opt Control 24(1):6–8, 32
6. Welch G, Bishop G (1995) An introduction to the kalman filter. Technical Report TR 95–041 8(7):127–132
7. Mao X, Wu J (2017) Research on kalman filtering algorithm. Ship Electr Countermeasure 3:64–68
8. Sun Q (2011) Application status of improved kalman filter algorithm in state estimation of driving vehicles. Value Eng 19:51
9. Lindley DV, Smith AFM (1972) Bayes estimates for the linear model. J Roy Stat Soc 34 (1):1–41
10. 商娟叶 (2016) Research on Bayes estimation algorithm based on data fusion. Autom Instrum 2:118–120
11. Dempster AP (1967) Upper and lower probabilities induced by a multi-valued mapping. Annals Math Stat 38(2):325–339
12. Wang X (2007) Research on the synthetic rules of conflict evidence [D]. North University of China
13. Han L, Zhou F (2006) Knowledge fusion based on DS evidence theory and its application. J Beijing Univ Aeronaut Astronaut 32(1):49–65
14. Passino KM, Yurkovich S (2001) Fuzzy control. Tsinghua University Press

15. Dou Z (1995) Fuzzy logic control technology and its application. Beijing University of Aeronautics and Astronautics Press
16. Driankov D, Hellendoorn H, Reinfrank M (1997) An introduction to fuzzy control. 239 (2):160–160
17. Han L (2006) Artificial neural network. Beijing University of Posts and Telecommunications Press, Beijing
18. Simon H (1994) Neural network: a comprehensive foundation. Neural networks: a comprehensive foundation. Prentice Hall PTR, pp 71–80
19. Mao J et al (2011) Development and application of artificial neural network. Electron Des Eng 19(24):62–65
20. Hecht-Nielsen (2002) Theory of the backpropagation neural network. In: International 1989 joint conference on neural networks. IEEE, pp 445–445
21. Qin K (2002) Division commander soldiers. Application of artificial neural networks in the sensor data fusion. Sens Microsys 21 (10):49–51
22. 禤家裕 (2011) Research and application of multi-sensor data fusion technology in UAV navigation system [D]. South China University of Technology
23. Yan Y-S, Zhai W-B (2016) Data fusion application technology in ship unmanned technology. Ship Sci Technol 20:1–3
24. https://blog.csdn.net/p%23onzq/article/details/79257269
25. Zhang L (2014) Design of unmanned ship information module based on information fusion technology [D]. Wuhan University of Technology
26. An J, Wen G, Lu Y et al (2009) Multi-sensor data fusion method for vehicle autonomous navigation. Automot Eng 31(7):640–645

# Chapter 7
# Positioning and Navigation Technology

## 7.1 Introduction

With globalisation and the rapid technological development, today navigation and positioning technology plays an increasingly important role in our daily work and life. It is widely used in modern advance technological fields particularly in the sea, road and air. In the unmanned autonomous mobile system, the navigation technology mainly provides information such as direction, position, speed and time for the motion carrier. It is used to determine the geographical location of the motion carrier itself, which is the basis and support for the next path planning and mission planning.

With the development of science and technology, people's requirements for navigation and positioning technology are getting much higher. In the military, navigation positioning technology is an important national information infrastructure. Strategic installations for civilian usage, in road vehicles, ships, aircraft and other transportation equipment rely on navigation positioning technology which encompass huge economic and social benefits. With the development of unmanned autonomous mobile systems, positioning and navigation technology has become a hot research topic.

## 7.2 Overview

Currently, the commonly used navigation positioning technology in the unmanned system could be divided into the following four categories:

(1) Relative positioning: It mainly relies on inertial sensors such as odometers and gyroscopes to determine the current position of the unmanned vehicle by measuring the displacement of the unmanned vehicle relative to the initial position;



X. Bi, *Environmental Perception Technology for Unmanned Systems*, Unmanned System Technologies, https://doi.org/10.1007/978-981-15-8093-2_7

(2) Absolute positioning: It mainly uses the navigation beacons for map matching or relying on Global Navigation Satellite System (GNSS) for positioning;
(3) Combined positioning: It combines relative positioning and absolute positioning. The combined positioning scheme generally has GNSS and track estimation, or GNSS and track estimation and map matching; and
(4) Simultaneous Localization and Mapping (SLAM).

Satellite navigation and positioning technology refers to the use of the satellite navigation and positioning system which provides the location, speed and time information to facilitate the positioning, navigation, supervision and management of the various targets. The global navigation satellite system occupies an important position in the field of navigation due to its global, all-weather and real-time nature. However, the satellite navigation technology still has many shortcomings at the present. The most obvious of which is the poor autonomy, easy interference, and low update rate. Therefore, it is highly undesirable to use the GNSS alone when high precision is required.

Inertial navigation technology utilizes the gyroscope and the accelerometer as sensitive devices. The navigational coordinate system is established based on the gyroscope's output and the velocity and position of the carrier in the navigation coordinate system is calculated based on the accelerometer's output. Inertial navigation technology is a technology developed in the early twentieth century and is a typical fully autonomous navigation method. Since the inertial navigation component is small and is low in cost, and thus the inertial navigation technology has been widely used in various fields. As the effective information is acquired from an accelerometer and a gyroscopic instrument, it is therefore not dependent on any external information. Additionally, it does not emit electromagnetic waves and has a strong independence. Furthermore, the frequency of updating the parameters of the inertial navigation device is high, and so the system has a better real-time performance and fast response speed. As such, the system can provide the carrier all the required navigation data. However, inertial navigation also has its disadvantages. As the inertial navigation obtains the attitude, position and velocity through the integral operation of the acceleration and angular velocity, and thus, the accuracy of the navigation is limited by the accuracy of the accelerometer and the gyroscope. The cost and mass of the high precision inertial navigation component will inevitably increase. Hence, this will greatly restrict the application of the inertial navigation. Furthermore, the navigation error will accumulate with the integral operation, and hence navigation for a long period of time will develop a very large drift.

Since each single navigation system has its own limitations, and thus integrated navigation has become a hot research topic. The integrated navigation system is a combination of the various navigation systems which fully utilizes the effective information from each sensor. This allows the various systems to complement one another's shortcomings, constituting a high-precision and highly reliable navigation system. The integration of the inertial navigation and the satellite navigation is a technology with vast prospects and is undergoing rapid development.

Nonetheless, there are many problems that still need to be solved to achieve true autonomous navigation. This is because when the unmanned mobile system is in an unknown scenario, it does not have pre-known maps of the surroundings and hence is unable to achieve its own positioning. Additionally, in the absence of the satellite signals or inaccurate positioning, it is also difficult for the autonomous mobile systems to rely on the environmental maps for navigation and positioning. The unmanned autonomous mobile system starts to move from an unknown location in an unknown environment, and locates itself according to the position estimation and the map during the movement process. At the same time, it builds an incremental map based on its own positioning to realize the technology of autonomous positioning and navigation called Simultaneous Positioning and Map Creation (SLAM). In SLAM technology, the process of self-positioning of mobile robots is carried out concurrently as the process of creating the scenario maps. Therefore, the basic theories and solutions for the in-depth study of SLAM-related problems constitutes great significance in research and application value in realizing the autonomous navigation of the unmanned autonomous mobile systems.

## 7.3 Global Navigation Satellite System

Currently, the most prominent satellite navigation system is the United States' Global Positioning System (GPS). However, since the GPS in the United States only provides accurate positioning signals to the local military, but low-precision signals to other users after added interference, countries and organizations worldwide had begun to establish their own satellite navigation and positioning systems in order to break away from monopolization and the dependence on the United States' GPS. For example, Russia's Global Orbiting Navigation Satellite System (GLONASS), Europe's Galileo Satellite Navigation System (GALILEO) and China's Beidou Satellite Navigation System (BDS). The following takes GPS as an example to introduce the basic principles of the satellite navigation.

### *7.3.1 Composition of GPS System*

The GPS system consists of three parts, namely, the space segment, the control section and the user segment.

The GPS space is composed of 24 satellites, namely 21 working satellites and 3 in-orbit spare satellites, as shown in Fig. 7.1. The 24 satellites are distributed around the earth in six orbits with an inclination of 55°. The orbital planes are 60° apart and the average orbit is 20,200 km. The satellite's operating cycle (i.e. the time around the Earth) is about 12 stellar time (11 h and 58 min). The main role is to send satellite signals for navigation and positioning.

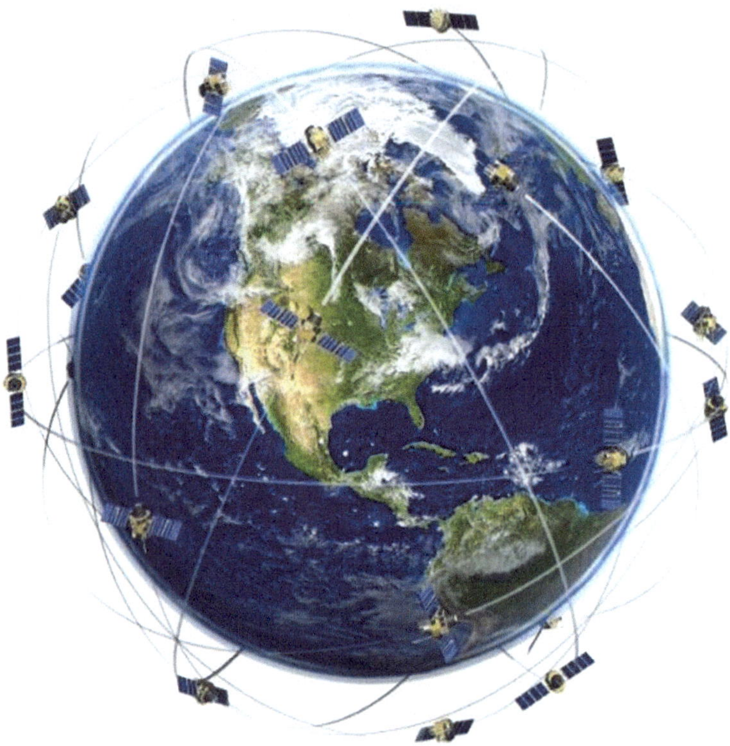

**Fig. 7.1** GPS satellite distribution map

GPS control segment consists of 1 main control station, 5 testing stations and 3 injection stations. The role is to monitor and control the satellite operations, calculate satellite ephemeris (navigation message) and maintain system time.

The main control station: Collects the satellite data from each monitoring station, calculates the satellite ephemeris and clock correction parameters, and injects satellites through the injection station. It issues instructions to the satellites, controls the satellites, and dispatches the standby satellites when the satellites fail.

Monitoring station: Receive satellite signals, detect satellite operating status, collect weather data and transmit this information to the master station.

Injection station: Inject satellite ephemeris and clock correction parameters calculated by the master station into the satellite (Fig. 7.2).

The GPS user equipment section contains a GPS receiver and related equipment. The GPS receiver is mainly composed of a GPS chip, such as a carrier, a shipborne GPS navigator, a mobile device with a built-in GPS function, and a GPS surveying device, all of which belong to a GPS user device.

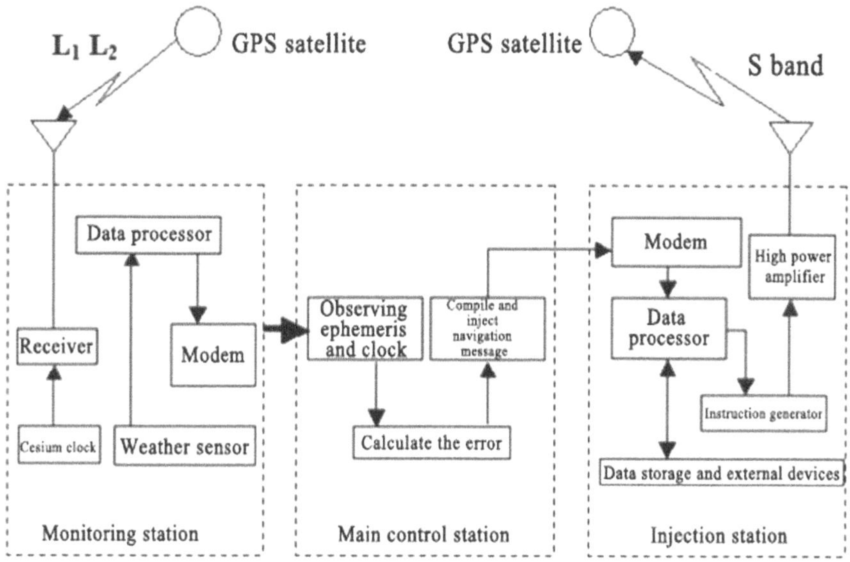


**Fig. 7.2** Framework of ground control segment

## 7.3.2 *Principle of GPS Positioning*

GPS users base on the distance measurements of multiple satellites and the concept of geometry, with the known location of the satellites, to calculate the specific location of the user. The GPS satellite transmits two carrier signals, L1 and L2. The L1 frequency is 154 f (f is the base frequency: 10.23 MHz) and the wavelength is 19.03 cm. The coarse acquisition code C/A code and the fine code P code are modulated and both modulate the data of 50 bps. These data are known as navigation message which provide users with essential information for navigation. For terrestrial users, the C/A code rate is f/10 and the P code rate is 10 f. The general positioning service (SPS) measures the distance from the user to the satellite according to the C/A code, and its positioning accuracy is about 10 m which mainly serves ordinary users. The Precision Positioning Service (PSS) measures the distance from the user to the satellite based on the P code. The positioning accuracy is about 1 m, which is mainly used by the US military.

The GPS navigation satellite carries a high-precision atomic clock. The atomic clock controls the GPS navigation satellite to continuously transmit navigation information to broadcast at a certain timing. The navigation broadcast contains time information, and the user can calculate the time when the navigation message leaves the satellite according to the received navigation broadcast. As long as the time of receiving the navigation message is recorded, it is convenient to calculate the propagation time of the navigation message from the satellite to the user. The distance from the user to the satellite is obtained by multiplying the propagation time with the speed of light. At the same time, the navigation message contains the ephemeris information of each satellite, and the exact position of the satellite can be obtained according to the ephemeris. In theory, the user can determine the user's location by receiving the navigation message of three satellites at the same time. Three satellites are used to determine three distances and three spherical surfaces. The spherical center determines the satellite position and the satellite spherical radius is the distance from the user to the satellite. The user must be located on the intersection of three spherical. The equations are:

$$\begin{cases} \sqrt{(x_1-x)^2+(y_1-y)^2+(z_1-z)^2}=d_1 \\ \sqrt{(x_2-x)^2+(y_2-y)^2+(z_2-z)^2}=d_2 \\ \sqrt{(x_3-x)^2+(y_3-y)^2+(z_3-z)^2}=d_3 \end{cases} \quad (7.1)$$

where $(x_i, y_i, z_i)$ is the spatial coordinates of the $i$th satellite, $(x, y, z)$ is the coordinates of the user, and $d_i$ is the distance from the $i$th satellite to the user (Fig. 7.3).

In practice, the clock at which the user records the reception time of the navigation message is usually quartz with lower precision which is not synchronized with the GPS system, and the time difference of t is unknown. Even if the atomic clock carried by the satellite does not synchronize with the GPS system, the actual master station will periodically correct the satellite clock, and the navigation message has the clock difference t of each satellite. Hence, in engineering calculations, it is necessary to introduce a fourth navigation satellite to determine the

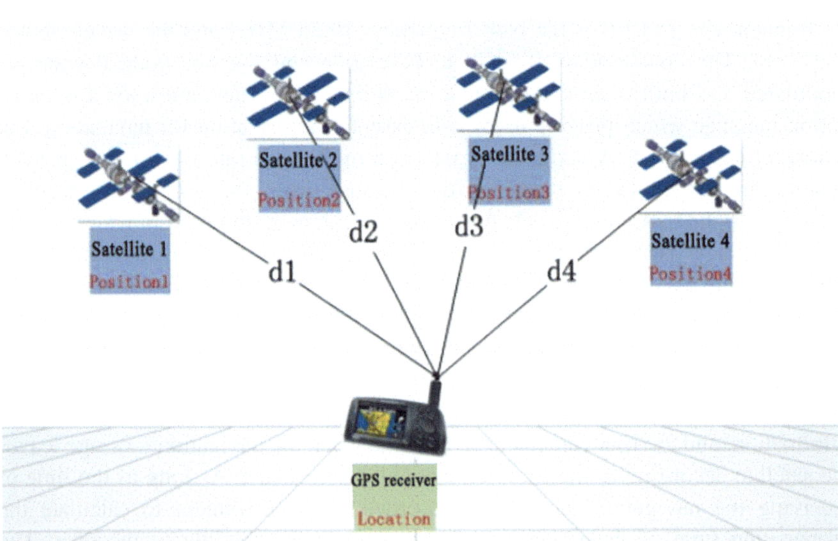


**Fig. 7.3** Diagram of GPS positioning

GPS receiver clock and the GPS system time difference, t. The equations are shown as follow, where c is the speed of light:

$$\begin{cases} \sqrt{(x_1 - x)^2 + (y_1 - y)^2 + (z_1 - z)^2} + c(t_1 - t) = d_1 \\ \sqrt{(x_2 - x)^2 + (y_2 - y)^2 + (z_2 - z)^2} + c(t_2 - t) = d_2 \\ \sqrt{(x_3 - x)^2 + (y_3 - y)^2 + (z_3 - z)^2} + c(t_3 - t) = d_3 \\ \sqrt{(x_4 - x)^2 + (y_4 - y)^2 + (z_4 - z)^2} + c(t_4 - t) = d_4 \end{cases} \tag{7.2}$$

### 7.3.3 Differential GPS Positioning Technology

In order to improve the accuracy of the positioning, differential GPS technology is usually used to locate the vehicle. As the differential GPS positioning method can completely eliminate the errors common to multiple receivers, such as satellite clock errors and ephemeris errors, it can eliminate most of the delay errors such as ionosphere and tropospheric propagation. Thus, significantly improve the positioning precision as compared to single-point positioning. The differential GPS system consists of a base station, a data transmission device, and a mobile station. According to the different ways of transmitting information by the differential GPS reference station, the GPS differential positioning can be divided into the following three categories: position differential, pseudo-range differential and carrier-phase differential. The three types of differential work in a similar manner, that is, the base station sends a correction number to the mobile station, and the mobile station uses the correction number to correct its own measurement result, thereby obtaining accurate positioning results. The difference is that the specific content of the sent correction number and the differential positioning accuracy are different (Fig. 7.4).

(1) Position differential: It is the simplest differential method for all GPS receivers. The position difference requires that the base station and the mobile station observe exactly the same set of satellites. The correction number is the position correction number, that is, the receiver on the base station observes the GPS satellite to determine the observation coordinates of the station. The difference between the known coordinates of the station and the observed coordinates is the position correction number.
(2) Pseudo-range differential: It is the most versatile technology. The correction number refers to the distance correction number. That is, it uses the reference station coordinates and the satellite ephemeris to calculate the distance between the station and the stars. The distance correction number is obtained by subtracting the observation distance from the calculated distance.
(3) Carrier-phase differential: It is also known as the RTK technology which is based on the real-time processing of the carrier phase of two stations. It can

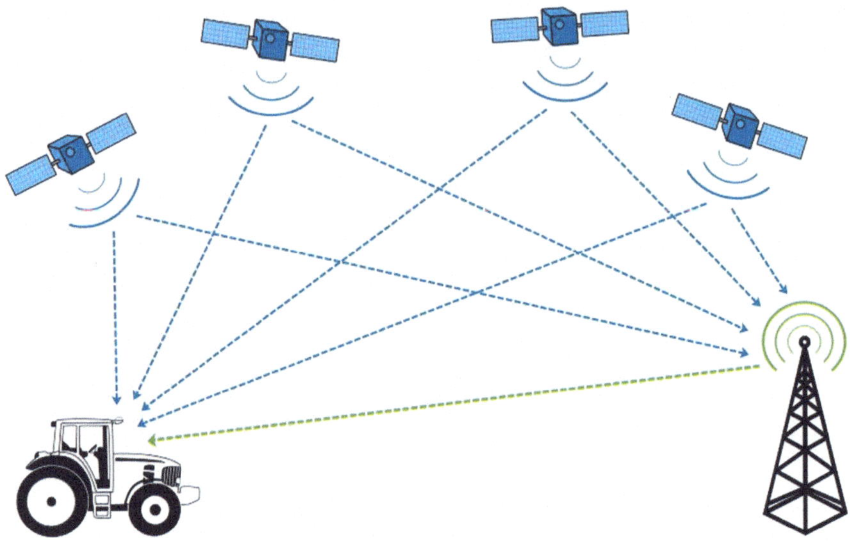

**Fig. 7.4** Diagram of differential GPS positioning

provide real-time, 3D coordinates of the station within centimeter accuracy. The carrier-phase differential is further divided into the original phase observation differential and the phase correction differential. The former is similar to the static measurement, which transmits the carrier phase in real-time from one station to another to jointly solve the baseline components. This type of differential technique can achieve positioning within centimeter accuracy, but there exists a critical problem to solve for the phase ambiguity in real-time. The latter is to send the pseudo-range and phase correction number by the base station so that the user can use the phase correction number to perform the point calculation. This method is also called the quasi-carrier phase differential, and the precision can reach the decimeter level. The quasi-carrier phase differential transmits the carrier-phase correction number instead of the carrier-phase measurements. This requires a small dynamic range and a narrow frontal band.

## 7.4 Inertial Navigation System

Inertial navigation system is theoretically based on Newton's laws. The acceleration vector with respect to the selected coordinates is measured by the accelerometer. The corresponding displacement is obtained during the second integration. By using the gyroscope's rotational angular velocity, the rotation angle is obtained after the first integration. The above process is iterated many times to calculate the real-time position.

As there is no signal communication with the outside of the carrier, the inertial navigation system has complete autonomy. The inertial navigation system has strong adaptability and no requirements for the working environment. The system can realize navigation and positioning on a global scale without any external information intake. It is precisely because of this series of features that it has a very wide range of applications in aerospace, aviation, navigation and other fields.

## 7.4.1 Composition of the Inertial Navigation System

Inertial navigation systems typically consist of the inertial measurement devices, computers, control displays, etc. The inertial measurement device includes an accelerometer and a gyroscope, also known as an inertial navigation combination. A gyroscope usually refers to a rotor that is mounted at a high speed in a gimbal. The rotor can be precessed around an axis or two axes perpendicular to the axis of rotation. The former is called a single-degree-of-freedom gyroscope, and the latter is called a two-degree-of-freedom gyroscope. Single-degree-of-freedom gyroscope is sensitive to the angular velocity while the two-degree-of-freedom gyroscope is sensitive to the angular displacement. The gyroscope has a fixed axis and precessibility. These characteristics are used to produce a rate gyro with a sensitive angular velocity and a position gyro with a sensitive angular deviation. Accelerometers are one of the core components of an inertial navigation system. By relying on its measurement of force, the inertial navigation system is able to determine the position and speed of the carrier and complete the task of generating a tracking signal.

In the inertial navigation system, three degrees of freedom gyros are used to measure the three rotational motions of the carrier. Three accelerometers are used to measure the acceleration of the three translational motions of the carrier. The computer calculates the speed and position data of the carrier based on the measured acceleration signal. The control display exhibits various navigation parameters to implement the navigation and positioning functions.

## 7.4.2 Classification of the Inertial Navigation System

1. **Platform Inertial Navigation**

The inertial navigation system with physical stability platform is called the platform inertial navigation system. The stable platform is the core part of the platform inertial navigation system. The accelerometer used to measure the acceleration of the carrier is mounted on the inertial platform. The system utilizes the precession of the gyroscope and applies torque control to offset the rotation angle of the gyroscope to maintain relative stability with the inertia space. The specified navigation coordinate system is always tracked, which avoids the influence of the carrier motion on the acceleration measurement and provides a measurement reference for the entire system.

The platform in the platform inertial navigation can avoid the influence of the carrier motion on the inertial components. The angle sensor on the frame can directly measure the attitude angle used for navigation estimation. At present, the platform inertial navigation system has developed to a very advanced level. However, its cost is high. The maintenance cost involved in the later stage is expensive and its reliability is not guaranteed since the servo system is adopted.

2. **Strapdown Inertial Navigation**

The strapdown inertial navigation requires for the gyroscope and the accelerometer to be directly mounted on the carrier. The gyroscope and the accelerometer are respectively used to sense the angular velocity and linear acceleration information of the carrier. At the time of installation, the two inertial elements are mounted one in each direction of the three-dimensional coordinate system, and that the input axes of the same type of elements are orthogonal to each other. The attitude matrix is calculated from the angular velocity collected by the gyroscope, and functions as a "mathematical platform". In the attitude matrix, information such as the heading and attitude of the carrier can be extracted. The output of the accelerometer in the carrier coordinates is the left-multiplying transfer matrix. The acceleration information in the corresponding navigation coordinate system is obtained, and then the transferred acceleration is solved.

The strapdown inertial navigation system eliminates the inertia platform which greatly reduces the quality, size and cost of the entire system. Its sensitive components are easier to install, repair or replace. However, the inertial component is directly attached to the carrier, and hence, the carrier vibration will impact the inertial component. As a result, the accuracy of the system is affected. Therefore, it is necessary to formulate a corresponding error compensation scheme or use a component with better performance.

## *7.4.3 Characteristics of Inertial Navigation System*

The advantages of the inertial navigation systems mainly include the following aspects: (1) the inertial navigation system is neither dependent on any external information nor on external radiation energy of autonomous systems. Hence, it has good concealment and is not affected by external electromagnetic interference; (2) it can function at anytime and anywhere globally whether in the air, on the ground or under the water; (3) it can provide the position, velocity, heading and attitude angle data, resulting in good continuity of the navigation information and low noise; and (4) high frequency of data update, and good short-term accuracy and stability.

Due to the impact by various factors, the inertial navigation system inevitably exists some disadvantages: (1) the navigation information system (INS) is generated by integral calculation, and hence, the positioning error will increase with time, resulting in poor long-term accuracy; (2) the initial calibration time prior to each use is much longer; and (3) the inertial navigation system equipment is more expensive.

### 7.4.4 Track Estimation Technology

In an inertial navigation system, trajectory estimation is performed to determine the position of the carrier. Trajectory estimation uses the momentary position of the carrier, heading and speed to estimate the current position. The position and the motion trajectory are being calculated according to the real-time measurement of the motion carrier's travel distance and heading. It is generally not affected by the external environment. However, since its own error is accumulated over time, it is not possible to maintain high precision for a long time when working alone.

Taking a two-dimensional plane as an example, assuming that the motion carrier is a mass point, the two-dimensional motion is performed on one plane, and the motion analysis is performed in a two-dimensional, right-angle plane coordinate system. The trajectory estimation adopts the absolute coordinate system, of which usually the axis Y points to the north and the horizontal axis X points to the east, as shown in Fig. 7.5.

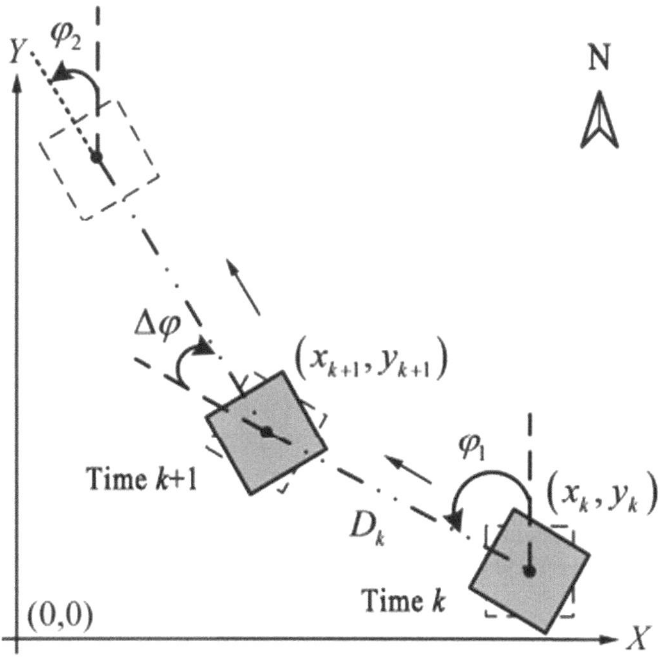


**Fig. 7.5** Trajectory estimation positioning method

By setting the position coordinate of the previous time as $(x_k, y_k)$, the current time's position coordinate is given as $(x_{k+1}, y_{k+1})$. The displacement of the adjacent time is $D_k$, the heading angle is $\varphi$, and the position of the current time can be obtained as follow:

$$\begin{cases} x_{k+1} = x_k + D_k \cdot \sin \varphi \\ y_{k+1} = y_k + D_k \cdot \cos \varphi \end{cases} \tag{7.3}$$

## 7.5 Integrated Navigation Positioning

An inertial navigation system is an autonomous navigation system that does not rely on external information and does not radiate energy externally. The basic working principle is based on Newton's law of mechanics. By measuring the acceleration of the carrier in the inertial reference frame, integrating it with time, and then transform it into the navigation coordinate system, the speed, yaw angle and position can be obtained. The inertial navigation system has the advantages of complete autonomy and could work under any weather and at any time. However, as the navigation information is generated through integration, and hence, the positioning error increases with time which results in poor long-term precision. Amongst the inertial navigation systems, the strapdown inertial navigation system (SINS) has a wider application. The global navigation satellite system (GNSS) is able to provide user with the 3-dimensional coordinate information, speed and time on Earth's terrestrial surface or any near ground spatial locations under any type of weather and at any time. However, under war conditions, GNSS positioning satellite may face restriction. The GNSS signals are susceptible to interference and the GNSS receiver data update rate is low. It is difficult to meet the requirements for overriding.

GNSS/SINS integrated navigation system overcomes the respective individual's shortcomings so that the combined navigation accuracy is higher than the accuracy of each independent system. The benefits of the integrated system include the ability for the inertial sensor to be calibrated for the inertial navigation system, to perform air alignment of the inertial navigation system, the stability of the height channel, etc., which can effectively improve the performance and accuracy of the inertial navigation system. In terms of global positioning system, the assistance from the inertial navigation system can improve the ability of the track satellites, which enhances the dynamic characteristics and anti-interference of the receiver. Therefore, the integration of the inertial navigation to establish an integrated system is an ideal navigation and positioning system project.

### 7.5.1 Loose Combination

In a loosely combined system, the GNSS and SINS work independently. The position and velocity information obtained by the GNSS and the rate information obtained by the IMU through SINS are combined by a filter. The obtained result is used to correct the SINS error so that the SINS can always maintain a high navigation precision. This combination mode is widely used because it is simple and easy to implement. The GNSS/SINS loose combination pattern diagram is shown in Fig. 7.6.

### 7.5.2 Tight Combination

The tight combination system utilizes the pseudorange and pseudorange rates of SINS and GNSS. GNSS can obtain pseudorange and pseudorange rate through the RF signal processor, GPS code ring and carrier tracking loop. The IMU solves the pseudorange and pseudorange rate by combining SINS with the GNSS satellite ephemeris. The SINS and GNSS pseudorange and pseudorange rate are directly sent into the combination filter and then correct using the SINS solver to obtain a higher accuracy navigation. Experiments had been conducted using this combination method theoretically and in engineering applications. The practical results show that the performance is better than the loose combination of the position and the speed, which effectively improves the observability of the combined system. The principle of the GNSS/SINS tight combination mode is shown in Fig. 7.7.

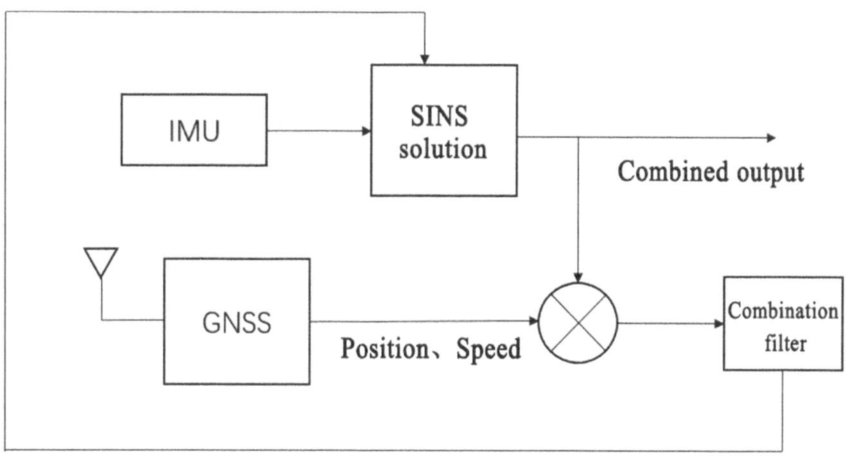


**Fig. 7.6** Concept of GNSS/SINS loose combination mode

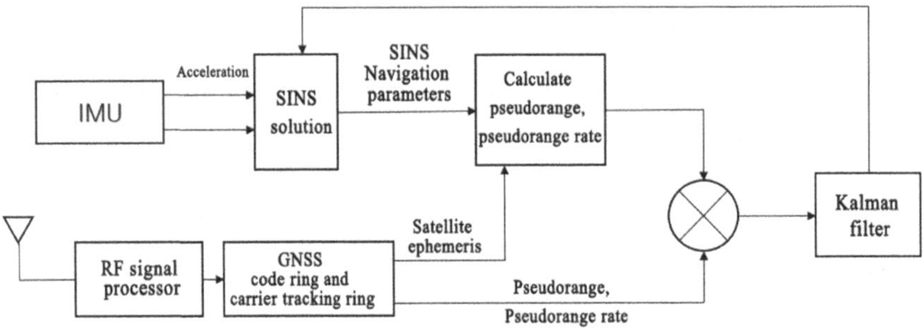


**Fig. 7.7** Principle of the GNSS/SINS tight combination mode

### *7.5.3 Ultra Tight Combination*

The ultra-tight combination mode not only combines the observed pseudorange and pseudorange rate parameters, but also feeds back the rate information of the SINS output to the tracking loop during the tracking loop. The workflow is closely similar to the tight combination. However, after the pseudorange and pseudorange rate are calculated by the SINS solver, the information is feedback to the GNSS code and the carrier tracking loop ring. This, not only enhances the tracking performance of the combined system for loop signals, but also reduces the occurrence of loss of lock. The principle of GNSS/SINS ultra-tight combination mode is shown in Fig. 7.8.

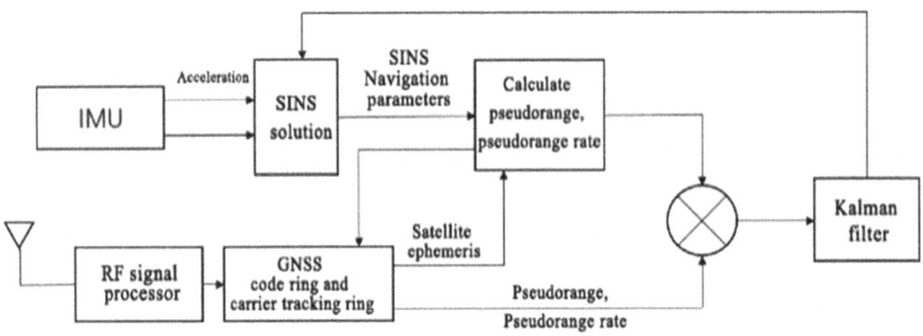


**Fig. 7.8** Principle of GNSS/SINS ultra-tight combination mode

## 7.6 Simultaneous Localization and Mapping (SLAM)

Simultaneous localization and map (SLAM) construction typically refers to a system of position and scenic map acquisition through the collection and computation of the data obtained from the various sensors that are mounted on the robots or the other forms of carrier. SLAM technology is critical for the action and interactive ability of the robots or other smart devices as it lays the foundation of knowing the location, the environment and how to act autonomously. It has a wide range of applications in autonomous vehicles, service robots, drones, AR/VR, etc. It can be said that any agent with certain mobility has some form of SLAM system.

In general, SLAM systems include multiple sensors and functional modules. According to the core functional modules, the current common robot SLAM system generally has two forms: laser radar-based SLAM (laser SLAM) and vision-based SLAM (Visual SLAM or VSLAM.

Laser SLAM was born out of the early range-based positioning methods (such as the ultrasound and infrared single-point ranging). The emergence and popularity of the lidar makes measurement faster and more accurate. Moreover, the information is more abundant. The object information collected by the laser radar presents a series of scattered points with accurate angle and distance information, which is called a point cloud. Generally, the laser SLAM system calculates the distance and attitude of the relative motion of the lidar by matching and comparing two point clouds at different times. Thus, completing the positioning of the robot itself. The lidar distance measurement is relatively accurate; the error model is simple; the operation is stable in an environment other than direct sunlight; and the processing of the point cloud is relatively easy. At the same time, the point cloud information itself contains direct geometric relationships, making the path planning and navigation of the robot intuitive. The theoretical research of laser SLAM is also relatively mature, and hence, the products are more abundant. Figure 7.9 shows the results of a map construction using a laser SLAM.

Vision SLAM uses a camera as a sensor that captures massive, redundant texture information from the environment and has superior scene recognition capabilities. The early visual SLAM was based on the filtering theory. Its nonlinear error model and huge computational complexity became its obstacles in practical application. In recent years, with the sparse nonlinear optimization theory (Bundle Adjustment) and the advancement of camera technology and computing performance, a large number of visual SLAM methods using cameras as sensors have emerged. Figure 7.10 shows the localization and map construction results of a visual SLAM.

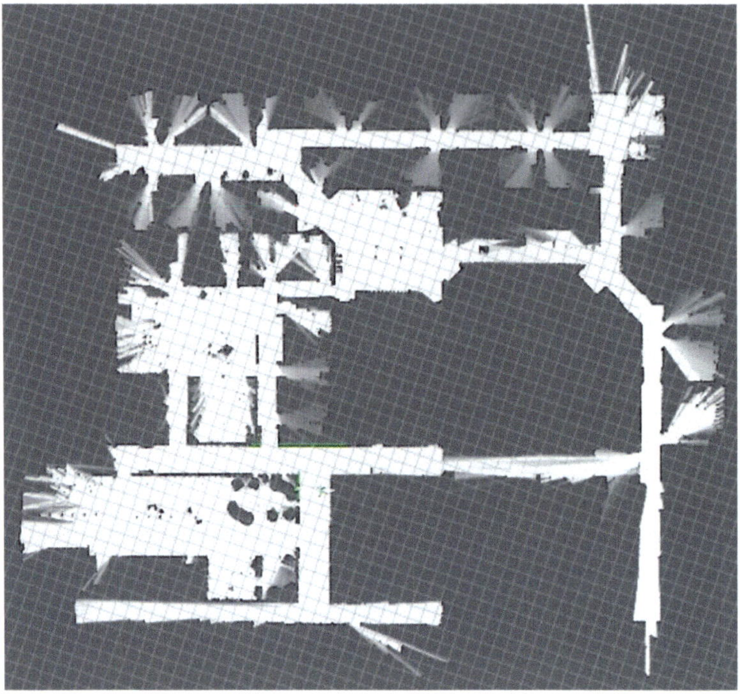

Fig. 7.9 Map construction of the laser SLAM

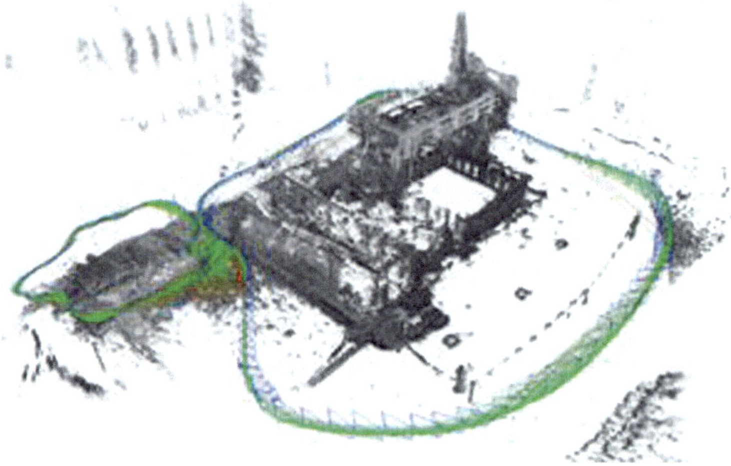

Fig. 7.10 Map construction of a visual SLAM

### 7.6.1 SLAM Implementation

The SLAM problem can be seen as a state estimation problem, which estimates the internal hidden state variables through the noisy measurement data. In general, there are currently two main options:

The first one is the filter method. In SLAM, the position state is used for filter estimation when Gaussian filtering or non-parametric filtering method based on Bayesian estimation is generally used. Parametric filtering methods include Kalman filtering (KF), extended Kalman filtering (EKF), information filtering (IF), extended information filtering (EIF), etc., while non-parametric filtering includes histogram filtering (HF) and particle filtering (PF), etc. The KF and IF require the system state equation to be linear, and the system noise and observation noise to be Gaussian noise. The EKF and EIF allow the system state equation to be nonlinear, but the noise is also required to be Gaussian. The particle filter is fitted by the sampled value. System noise, because of its wide adaptability to noise types, has been greatly developed and applied.

Another method is the nonlinear optimization method, which has been widely used in recent years. It uses state data collected at all times for state estimation and is considered to be superior to conventional filters. Hence, it has become the mainstream method of the current visual SLAM. The commonly used nonlinear optimization method is the graph optimization.

1. **SLAM Method Based on Kalman Filtering**

The Kalman filtering-based SLAM method uses the wide vectors including the unmanned system pose vector and the environmental eigenvector to represent the spatial environment. It describes the relationship between the unmanned system motion and the environmental characteristics as two nonlinear models, namely, the unmanned vehicle motion model and observation model. The unmanned system control signal is input into the system motion model to realize the motion of the unmanned system. The Kalman filtering algorithm realizes the prediction of the pose of the unmanned system according to the system model, and obtains the observation of the environmental characteristics according to the system observation model. The processing of data association matching is required between the predicted feature and the observed feature. The best matching feature is selected for updating the pose of the unmanned system, and the candidate matching feature is considered to be a new feature obtained for environmental observations, which is used for the widening of the map.

Kalman filtering assumes that the system is a linear system, but in reality, the motion model and the observation model of the unmanned system are both nonlinear, and so the EKF method is usually used. The EKF is expanded by a first-order Taylor to approximate the nonlinear model. The EKF-based SLAM can be summarized as a loop iterative estimation-correction process. That is, firstly the new position of the unmanned system is estimated by the motion model, and the observed environmental characteristics are estimated by the observation model.

Then, the error between the actual observation and the estimation observation is calculated. The integrated system covariance calculates the Kalman filter parameter K, and uses K to correct the position of the previously estimated unmanned system, before finally adds the newly observed environmental features to the map. The biggest drawback of the extended Kalman filtering is the assumption that the uncertainty in the system is consistent with the Gaussian distribution, and hence, it is helpless pertaining to the noise of other models in the system. Furthermore, the KF/EKF SLAM cannot handle the correlation problem, that is, the problem of data association. In severe cases, inaccurate data associations will cause the algorithm to diverge.

The critical problem facing the application of EKF for map creation and positioning is the real-time problem of localization. This is because of the presence of uncertainty in the pose and environmental characteristics of the external sensor information. The correlation metrics are necessary as the uncertainty in the poses are used for updating the feature map while the uncertainty in the environmental features are used to update the poses. The correlation between the driverless vehicles and the environmental features cannot be propagated independently. Therefore, in the process of updating the pose and environment features, it is necessary to calculate the covariance matrix corresponding to the cross-correlation between the unmanned system and the environmental features, and the environmental features versus the environmental features. Research to reduce computational complexity has focused on reducing secondary features, improving map representation and updating.

2. **Particle Filter-Based SLAM Method**

Particle filter localization uses a set of randomly weighted particles to obtain an approximation of the probability distribution, so it does not require that the noise must strictly follow the Gaussian distribution, and thus, can handle any part of the noise.

In the beginning of the particle filter positioning, an initial sample set needs to be set. Each particle in the set has a corresponding weight that represents the likelihood that the unmanned system is in the location of the particle. Then, by using the importance sampling technique, in each recursive process, the posterior probability of each particle in the sample is first predicted according to the motion model. The importance factor of each sample is calculated by the perceptual model, and the most probable sample is determined according to the importance factor. Thus, after multiple recursive iterations, samples with large weights may be selected multiple times, while particles with small weights are likely to be discarded. Thus, the larger error position is gradually replaced by the more likely position, that is, the precise position of the unmanned system is gradually obtained. Particle filtering was introduced into the research of SLAM, and a series of simultaneous positioning and map creation algorithms were proposed.

(1) FastSLAM Algorithm
In order to solve the problem that the EKF-SLAM algorithm is computationally complex in a wide range of real-life environments and is sensitive to failed data associations, the SLAM problem is decomposed into a positioning problem and a signpost estimation problem based on the pose estimation. The FastSLAM algorithm uses particle filtering for path estimation. The signpost pose estimation is implemented using the Kalman filtering, and each different roadmap uses an independent filter for the estimation of the landmark position under path estimation conditions.

(2) DP-SLAM Algorithm
The implementation of the FastSLAM algorithm requires pre-set artificial landmarks. In order to improve the practicability of the algorithm, the DP-SLAM algorithm is proposed, where no artificial landmarks need to be set in advance. The algorithm uses lidar as the environment perception device, and uses the particle filter to approximate the joint probability distribution of the environment map and the moving unmanned system pose, and performs synchronous positioning and map creation.

(3) SLAM algorithm Based on Rao-Blackwellized Particle Filter
In the Blackwellized algorithm, each particle represents a possible trajectory of the unmanned system, and each particle has its own global map. They echo the trajectory of the particle. Therefore, the algorithm can better approximate the joint probability density of the unmanned system pose and the environment map. However, when the number of particles and the size of the environment map increases, the algorithm will occupy a large amount of memory, and in the process of resampling Global maps, a lot of storage space is required for copying. Building on the RB filtering SLAM foundation, the prediction-based SLAM method (P-SLAM) is proposed. Through the prediction of the unexplored areas, and then compared with a known area, the synchronization localization and mapping is realized.

3. **SLAM Method Based on Information Filtering**

Like EKF, IF is a parametric filtering method that uses a Gaussian model for noise generated in motion and observation. The difference between IF and EKF lies in the form of the Gaussian noise. In Kalman filtering method, Gaussian noise is described by mean, $\mu$, and variance, $\sigma$. In the information filtering, normalized representation is used, which is described by information matrix, $\Omega = \sigma^{-1}$ and information vector $\xi = \sigma^{-1}\mu$. For arbitrary $x \sim N(\mu, \sigma)$, the probability expression is given as:

$$\begin{aligned} p(x) &= \det(2\pi\sigma)^{-\frac{1}{2}} \exp\{-\frac{1}{2}(x-\mu)^T\sigma^{-1}(x-\mu)\} \\ &= \det(2\pi\sigma)^{-\frac{1}{2}} \exp\{-\frac{1}{2}\mu^T\sigma^{-1}\mu\} \exp\{-\frac{1}{2}x^T\sigma^{-1}x + x^T\sigma^{-1}\mu\} \\ &= \eta \exp\{-\frac{1}{2}x^T\sigma^{-1}x + x^T\sigma^{-1}\mu\} \end{aligned} \tag{7.4}$$

where $\eta = \det(2\pi\sigma)^{-\frac{1}{2}} \exp\{-\frac{1}{2}\mu^T\sigma^{-1}\mu\}$ is a constant, and the normalized expression of the Gaussian model is:

$$p(x) = \eta \exp\{-\frac{1}{2}x^T \Omega x + x^T \xi\} \tag{7.5}$$

As the traditional EKF-SLAM method is proportional to the square of the number of features in computing time, which seriously affects the real-time performance of SLAM in large-scale environment, some scholars have proposed a SLAM method based on sparse extended information filter. By exploring the internal structure of the SLAM problem, the method uses the local mesh structure of the environmental features to represent the SLAM method, which has a fixed step size without the update time and feature quantity.

4. **SLAM Method Based on Graph Optimization**

Graph optimization is a way of representing optimization problems as graphs. It adopts the graph theory of configuring graph embodiment, which makes it more intuitive to represent the optimization problem in a particular form. A graph consists of a number of vertices, and the edges that connect the nodes. The vertices are used to represent the optimization variables and the edges are used to represent the error terms.

As shown in Fig. 7.11,the vertices and edges which include the moving body pose and map points constituting the vertices; the different pose moments between the apex and the pose; and the associated system-side configuration continue to accumulate to construct the entire map in the SLAM location model. The goal of the graph optimization is to adjust the respective vertex value such that the maxima is able to satisfy the constraints between the edges.

Once the graph is constructed, it is necessary to adjust the pose of the moving subject to satisfy the constraints of these edges. Thus, the SLAM problem based on graph optimization can be broken down into two tasks:

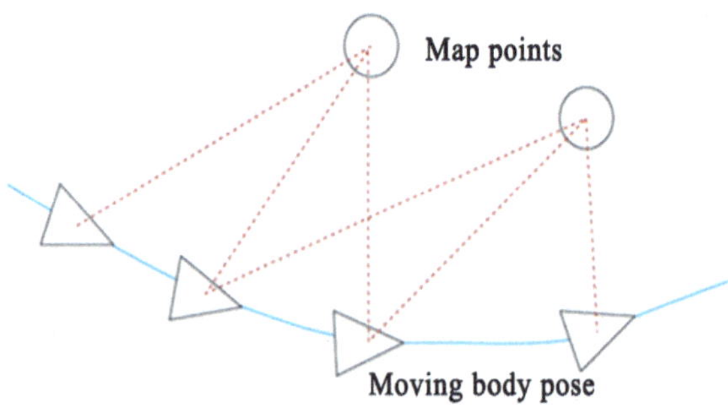


**Fig. 7.11** Graph optimization structure

1. Build the graph where the robot pose is set as the apex and the relationships between the poses as the edges. This step is often referred to as the front-end, which is usually the accumulation of the sensor information.
2. Optimize the graph and adjust the robot's pose vertices to satisfy the edge constraints as much as possible. This step is called the back-end.

### *7.6.2 Examples of SLAM Application*

SLAM is not only an important research direction in the field of robotics, but is also significant for unmanned systems which has received extensive attention and research. In comparison with the robots, the unmanned systems not only operate in a more complex environment, but also require real-time algorithms to ensure the execution of scheduled tasks. The SLAM method used in the example uses a lidar environment sensor to locate the unmanned vehicle through a histogram filtering technique on a two-dimensional plane and establish an Occupancy Grid Map (OGM). The whole process can be divided into two parts: the creation and update of the environment map and the positioning of the vehicle.

1. **Creation and Update of Environmental Maps**

There are several ways to describe the environment which include the raster maps, feature-based geometric information representation and topological maps. Due to the large environmental range and complex features in the experiment, it is not easy to extract a specific one or several features. Grid maps are easier to create and maintain. In addition, occupying the grid map can accommodate both the noise in the location information of the driverless vehicle and in the sensor signal. The OGM divided the environment into n grids with a fixed size $m_i$, $i \in [1, n]$. Here, the grid size is set to 0.2 m based on the requirements of the driver's obstacle avoidance. For each grid, m, by giving it a probability value of $p(m_i)(0<p(m_i)<1)$, it indicates the possibility of an obstacle at the grid location, which is given as:

$$p(m_i)\begin{cases} >0.5 & OCC \\ =0.5 & UNKNOWN \\ <0.5 & EMP \end{cases} \tag{7.6}$$

Among them, OCC means that there are obstacles in the grid, UNKNOWN means that the grid state is unknown, and EMP means that there are no obstacles in the grid. At the initial moment of the SLAM process, the initial state of all the grids in the map is UNKNOWN, indicating that the environment has not been explored.

During the SLAM process, the unmanned vehicle needs to be updated based on the results of the vehicle location and the map status data of the lidar. According to the Bayesian principle, the state of occupying the grid map at any time t can be calculated by the probability:

$$p(M|X_{1:t}, Z_{1:t}) = \frac{p(M|X_{1:t}, Z_{1:t-1}) \cdot p(Z_t|X_{1:t}, Z_{1:t-1}, M)}{p(Z_t|X_{1:t}, Z_{1:t-1})} \tag{7.7}$$

where t is the data acquisition time, X is the vehicle pose, and Z is the lidar measurement data. As the computation formula needs to record a large amount of historical data and the amount of data increases as the running time increases, therefore, it cannot meet the need for high-speed and real-time computation of the driverless vehicles. The above formula can be further simplified as:

$$p(M|X_{1:t}, Z_{1:t}) = \frac{S}{1+S}$$

$$S = \frac{p(M|Z_t, X_t)}{1 - p(M|Z_t, X_t)} \cdot \frac{p(M|Z_{1:t-1}, X_{1:t-1})}{1 - p(M|Z_{1:t-1}, X_{1:t-1})} \tag{7.8}$$

In the process of updating the map using the above formula, the pose of the vehicle $X_t$ is known. At the same time, the map status $p(M|Z_{1:t-1}, X_{1:t-1})$ at time $t-1$ is also known. The only unknown quantity, $p(M|Z_t, X_t)$, in the equation represents the measurement by the lidar in the current vehicle pose. As shown in Fig. 7.12, after the lidar data is projected into the grid map according to the vehicle position, $X_t$, the value of the corresponding grid, $p(M|Z_t, X_t)$, can be calculated.

2. **Vehicle Pose Calculation Based on Histogram Filtering**

In the process of calculating the position of the unmanned vehicle, the original positioning information of the vehicle is first obtained by the odometer that is installed on the driverless car. However, the positioning from the odometer is subject to certain errors, and it increases with the increase of driving time. Hence,

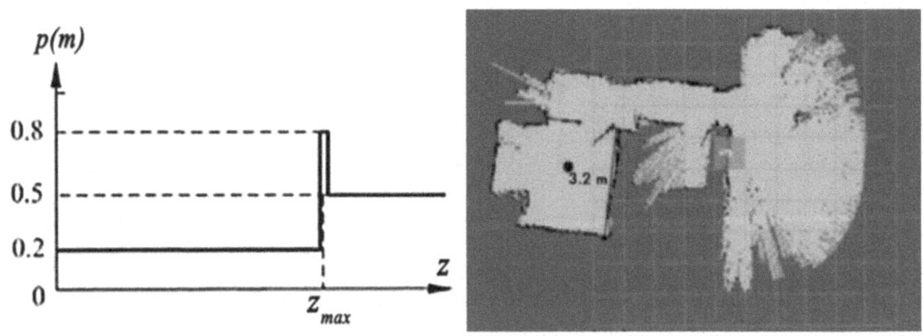


(a) Trans Sensor Model (b) Raster Map Created by Lidar

**Fig. 7.12** Sensor model and grid map. **a** Trans sensor model; **b** Raster map created by lidar

other measures are needed to correct it. Since the motion of the unmanned vehicle in the two-dimensional planar space is continuous, its position in the state space cannot be abruptly changed, so it is possible to estimate the deviation range of a vehicle positioned in the state space. This deviation range is referred as the state space deviation, as shown in Fig. 7.13. If the ideal position of the vehicle is at zero deviation position, the pose estimation range of the vehicle at the next moment can be obtained.

In the correction process of the original positioning information, the traversal operation is performed within the range of the positioning state space deviation, and in order to determine more accurate posture information, an evaluation mechanism is needed. The experiment used a voting mechanism. During the traversal process, the data of the lidar at this moment is projected onto the corresponding vehicle pose, and the measurement points of the lidar fall into different grids in the map. If the probability of the grid where the radar measurement point is located is greater than a predetermined threshold, the weight of the deviation position is increased accordingly. After comparing the probability values of the grids of all lidar measurement points, the weight of the deviation position can be expressed as:

$$score = \sum p(m_{t-1}^{hit_t^k}) \tag{7.9}$$

where, $m_{t-1}^{hit_t^k}$ represents the $k$th laser end point that corresponds to the grid in the global map, and $p(m_{t-1}^{hit_t^k})$ indicates the probability that the grid is occupied. After traversing all the deviation ranges, the deviation position with the largest weight is selected as the actual position of the vehicle and participates in the update process of occupying the grid map.

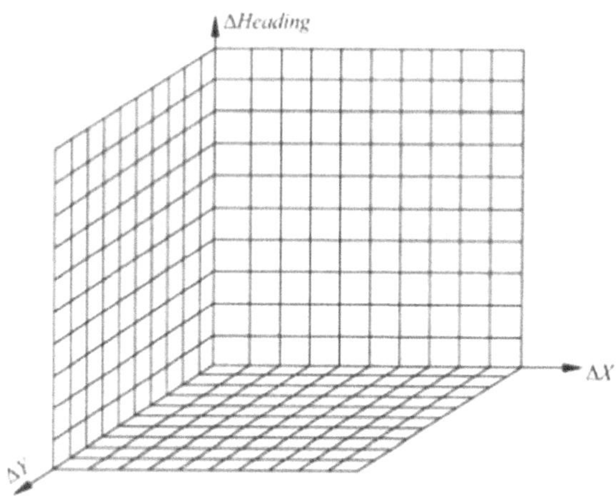


**Fig. 7.13** Discretized state space deviation

## 7.7 Navigation Application

### 7.7.1 Application in Unmanned Vehicle Positioning and Navigation

In the development of the automobile industry, the intelligent navigation of the automobile has always been the goal of the public, especially in the unmanned vehicles. With danger and uncertainty in mind, in order to reduce unnecessary casualties and the accident rate involving the unmanned vehicles, positioning and navigation technology has become an important development technology. High-precision driving positioning and navigation technology is one of the core technologies of autonomous driving. It is also the highlight of the autonomous driving layout. High precision here is defined as achieving centimeter-level error range, which is determined by the objective requirement for driving safety. In other words, if the error exceeds several tens of centimeters, collision between two vehicles may occur (Fig. 7.14).

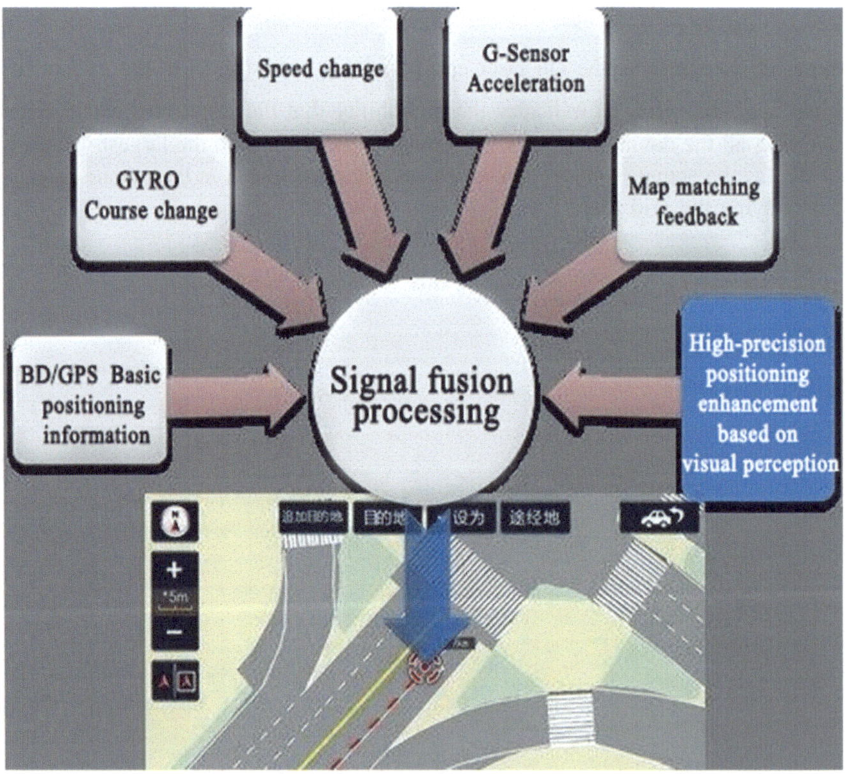


**Fig. 7.14** Unmanned vehicle positioning navigation system

Positioning navigation systems generally include GPS navigation, inertial navigation (INS), integrated navigation, etc. GPS and INS are currently more advanced navigation systems, each with its own advantages and disadvantages. GPS navigation has the advantages of global, all-weather, high-precision and three-dimensional positioning, but there are also shortcomings such as easy data loss and poor reliability. The INS rely solely on the carrier's own device for independent navigation and does not communicate with the outside world, so the data will not be disturbed. However, the INS error will accumulate over time and is not suitable for long-distance navigation. In terms of high-precision positioning and navigation technology, sensor fusion based on GPS and inertial sensors is an important positioning and navigation technology. Although the accuracy of GPS navigation can reach the level of centimeter range, when the self-driving car enters the GPS blind zone, bridges, tunnels, culverts or underground parking lots, it will be unable to locate themselves or signal shift, which will cause great security risks. It turns out that inertial navigation is also a necessary hardware device for automatic driving. Therefore, the two complement each other, which can effectively improve navigation accuracy and reliability.

In the integrated navigation system of an unmanned vehicle, the GPS receiver includes a base station and a mobile station. The mobile station is installed on the vehicle, and relatively high-precision positioning data (in the order of 10 cm) is obtained by the carrier difference. The inertial navigation system is fixed on the vehicle body and can provide (x, y, z) three-phase acceleration and direction angle when the vehicle is running. According to this, the position of the next time point can be inferred according to the current point, and the position of the inertial navigation system can be estimated. GPS-inferred data is fused by weight ratio to obtain a relatively accurate position. Subsequently, through the other environmental sensing sensors that are installed on the unmanned vehicle, the surrounding object information is scanned in real time, the data point information of the surrounding obstacles is acquired, and the contour of the surrounding obstacles is modelled by processing the data points, thereby forming a basis for obstacle avoidance.

Since the requirements for reliability and safety of the unmanned driving are very high, the positioning based on GPS and inertial navigation is not the only positioning method for unmanned driving. In reality, LiDAR point cloud and high-precision map matching, and vision are also used. Positioning methods such as odometer calculations allow various positioning methods to correct each other to achieve more accurate results.

### *7.7.2 Application in Unmanned Surface Vehicle Positioning and Navigation*

With the increasing demand for applications such as marine development and oceanographic surveys, the unmanned underwater vehicle (UUV) has been in marine survey and stereoscopic monitoring with its unmanned, cable-free and autonomous operational capabilities. The field is widely used. However, the precondition for UUV to achieve autonomous underwater operation is that it must have the ability of autonomous navigation. Traditional navigation technology usually combines Doppler velocimeter with inertial navigation system. Based on the dead reckoning algorithm, this method has certain limitation, that is, the error accumulation will increase with time. UUVs currently using SLAM technology can work on unknown terrain and be able to perform longer duration tasks. The potential advantages of the SLAM algorithm make it a booming stage in UUV navigation and positioning applications (Fig. 7.15).

Over the past two decades, UUV's positioning and navigation problems have attracted a large number of research groups, especially those working on the positioning navigation and the environmental perception research, which had proposed that the UUV to perceive a map of the unknown environment while determining their current location. This process only requires UUV sensors to sense the environment and its state, so it can be called an autonomous system.

The University of Sydney has conducted a detailed study of underwater SLAMs. They use mechanical scanning sonar as a sensor for their perceptual map features. The previous work was carried out in a swimming pool with artificial road signs with known locations. Point features are used in the final map to represent these observed targets. This setup provides a SLAM test method called the Geometric Projection Filter (GPF), which estimates the relationship between the independent landmarks rather than their actual position in the geodetic coordinate system. Subsequently, in the real-life, natural environment of the Sydney coast, a new experiment was conducted. This time, the classic EKF algorithm for random maps

**Fig. 7.15** Autonomous unmanned underwater vehicle

is the core of the SLAM system. Artificial landmarks are placed in this area to produce reliable point features. Although some natural landmarks are also detected, they are almost unstable and cannot be used for map creation. The SLAM algorithm was then studied, that combines the sonar and the visual system to derive an estimate of the motion of the craft. The data for this experiment was obtained when the Oberon aircraft was probed on the Great Barrier Reef in Australia. The camera mounted on the lower part of the aircraft is able to obtain a clear image of the coral reef while the aircraft is in operation. At the same time, the scanning sonar with the pen beam mounted on the camera can provide a series of distance measurements, and the features with larger differences can be marked in the image in the observable area of the sonar. Although this method reduces the dimensionality of the terrain model, it has a good improvement in the consistency of the convex and coral structures of the model (Fig. 7.16).

### *7.7.3 Application in Unmanned Aerial Vehicle Positioning and Navigation*

The unmanned aerial vehicle (UAV) is a drone that uses the radio remote control equipment and its own program control device. The drone consists of an aircraft, a control station, a communication link, a launching recovery device, etc. It has many advantages such as high manoeuvrability, high image resolution, small size, convenient portability, low cost, and no risk of casualties. UAVs are characterized by

**Fig. 7.16** UUV barrier reef detection

the ability to detect specific information in specific areas, and thus requiring the drones to be accurately positioned. At present, the UAV adopts navigation methods including inertial navigation, satellite navigation, Doppler navigation, visual navigation terrain-assisted navigation, and geomagnetic navigation. Amongst all, the combination of GPS and inertial navigation makes the navigation accuracy of the drone higher (Fig. 7.17).

The most significant combination of the GPS and the INS is the Kalman filter. The principle is to establish the state equation of the integrated navigation system based on the inertial system error equation, and establish the measurement equation of the combined system based on the navigation system error. The linear Kalman filter is used to provide the virtual distance and distance difference measurements required for the minimum variance estimation of the inertial system error. Then, the estimated values of these errors are used to estimate the error margin between the GPS measured value and the INS value. The results are modified in a feedback manner to reduce navigation errors. In addition, the calibrated inertial navigation system can provide navigation information to assist the system in improving its performance and reliability.

The combined navigation of the GPS and the INS can conduct over-the-horizon remote control of the UAV fight and the program-controlled autonomous flight. The over-the-horizon remote drone exceeds the scope of ground station observation and measurement and control. Hence, it is necessary to adopt an autonomous navigation method, that is, an integrated navigation system of the onboard navigation GPS/INS. The ground remote control station personnel obtains the information such as the attitude, azimuth, distance, speed and altitude of the drone in real-time through the integrated navigation system, and then, tracks, locates and controls the information. When the drone is found to deviate from the scheduled route, the attitude is

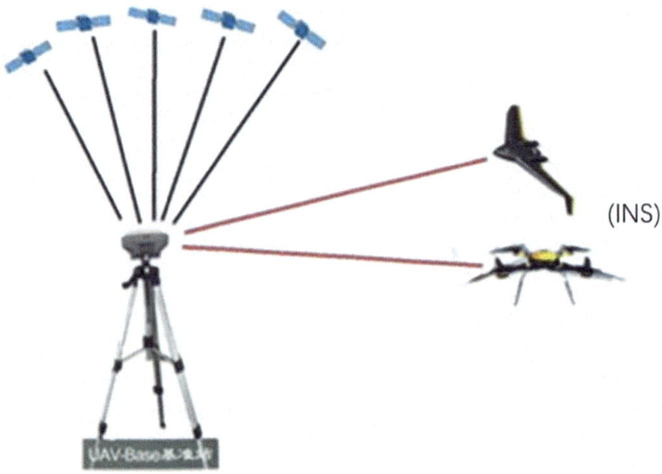


**Fig. 7.17** UAV combined positioning navigation

deviated and other flight conditions such as the flight state can be adjusted in time. The program-controlled UAV relies on the onboard flight control system and the onboard equipment to achieve flight control in accordance with the procedures and along predetermined routes. In the course of flight, various types of sensors, including GPS/INS integrated navigation system, are used to obtain the direction, position, altitude, speed and other information of the drone in real-time. After the calculation of the parameters by the flight control management computer, the activities are controlled by the autopilot such as the wing surface, throttle adjustment, attitude correction of the drone, in time tracking, and operating the flight according to the pre-set mission profile. The unique advantage of the GPS is undeniable. Traces of GPS applications can be observed everywhere whether in the military or civilian aspect. Although there are many advantages of GPS, there are also some shortcomings. The low precision of the short-term positioning requires the GPS to combine with other navigation methods to realise the full potential of the respective advantages. In the development of the UAV navigation system, the combination of the GPS and the INS is undoubtedly a perfect complement to each other, solving various problems in the navigation of some UAVs. However, the problem will continue to appear as the demand grows. Therefore, the application of GPS in the UAV navigation still needs to be further explored and discussed.

## Bibliography


1. Le Y (2006) Research on INS/GPS/PLS integrated navigation and positioning system [D]. Hehai University
2. Xu Z (2017) Research on vehicle integrated navigation technology based on MEMS IMU [D]. Wuhan University
3. Wang Y, Jin X, Hua H et al (2016) SINS/GPS research and analysis of integrated navigation technology. Electron World 22:53–54
4. Dong X (1998) GPS/INS integrated navigation positioning and its application. National University of Defense Technology Press
5. Yuan X (1993) Navigation system. Aviation Industry Press
6. Zhou Z, Yi J (1992) Principles and applications of GPS satellite measurement. Surveying and Mapping Press
7. https://blog.csdn.net/lqxandroid2012/article/details/47396787
8. Wang G (1999) Carrier phase differential GPS positioning technology. Surveying Mapp Eng (1):12–17
9. Zhao Y, Shi W, Ai M (2016) Indoor pedestrian track estimation/ultrasonic combined positioning fusion algorithm. J Central South University Nat Sci 47(5):1588–1598
10. https://baijiahao.baidu.com/s?id=1582920932965150131&wfr=spider&for=pc
12. Chen H (2014) Introduction to driverless vehicles. Beijing Institute of Technology Press
13. Kundra L, Ekler P (2013) The summary of indoor navigation possibilities considering mobile environment. In: European regional conference on the engineering of computer based systems. IEEE, Budapest, pp 165–166
14. Li C (2012) Research on navigation and positioning and environment sensing technology of unmanned underwater vehicles [D]. Harbin Engineering University
15. Qinglu Ma (1999) Vehicle positioning and navigation system. Central South University Press, Changsha

16. Cheng Y (2007) Research on simultaneous localization and map construction algorithm based on extended kalman filter [D]. Ocean University of China
17. Qu J (2009) Research on environment perception and terrain modeling method of sonar-based UUV [D]. Harbin Engineering University
18. Ford HJ (2012) Shared autonomous taxis: implement an efficient alternative to automobile dependency [D]. Princeton University
19. Wu Z (2017) Application of GPS global positioning system in UAV navigation system. SME Manage Technol (Mid-Term Publication) 3:110–111

# Chapter 8
# Autonomous Path Planning

## 8.1 Introduction

Path Planning refers to the search for the most optimal path that is subjected to a certain optimization criteria or criterion (shortest walking path, shortest travel time, etc.) and could avoid obstacles from an initial state to a target state. Path planning is one of the key technologies of unmanned systems. Almost every task implementation involves path planning. At the same time, path planning involves many complex technologies and operations such as environmental model building, lane changing, cornering, intersection operations, and most importantly, obstacle avoidance. Path planning algorithms are also widely adopted in many other fields of application, not just in the field of the mobile robots. Applications in the advanced technology fields include the UAV's obstacle-avoidance flight, the cruise missile avoidance radar search, the anti-bounce attack, and the completion of the blasting tasks. Applications in daily life include the GPS navigation, the GIS-based road planning, the urban road network planning and navigation. Applications in the field of decision management include the vehicle routing problem (VRP) in vehicular logistics management, resource management and resource allocation problems, and routing problems in the field of communication technology. Fundamentally, any planning problem that can be topologically be represented as points, edges and a network can be solved by the path planning method.

The path planning problem is essentially a graph searching problem. At present, the path planning algorithm in the two-dimensional space can be divided into the conventional and the artificial intelligent algorithm. The conventional algorithm refers to an algorithm based on graph search, such as the artificial potential field, Dijkstra, A* and A* variants. Sampling-based algorithms include the RRT and PRM algorithms. Algorithms based on curve interpolation comprise of the Bezier curves, the gyroscopic lines, the polynomial curves, etc. Last but not least, the numerical optimization algorithms. Traditional algorithms are generally simpler, but are not conducive to use in complex environments where multi-objective



X. Bi, *Environmental Perception Technology for Unmanned Systems*, Unmanned System Technologies, https://doi.org/10.1007/978-981-15-8093-2_8

optimization is required. Artificial intelligent algorithms include ant colony algorithm, genetic algorithm, neuron algorithm, evolutionary algorithm, etc. Such algorithms have a high search efficiency but long search time. The design of the algorithm forms the core of the path planning. Great progress has been achieved from the use of the traditional algorithms to the artificial intelligent algorithms that are developed in conjunction with the bionics. Different intelligent algorithms have different characteristics where the scope and area of application are different. Therefore, studying the path planning intelligent algorithm from the characteristics of the algorithm itself and its application is of great significance to the development of path planning technology.

## 8.2 Overview of Path Planning

Path planning technology is an important branch in the field of unmanned systems research. The optimal path planning problem in the unmanned system is based on one or some optimization criteria (such as the lowest work cost, the shortest walking route, the shortest walking time, etc.) to search for the most optimal route that could avoid obstacles from the origin to the destination. The existing path planning algorithm originates from the mobile robot. In this chapter, the path planning technology is further explained by introducing the unmanned vehicles. In order to quantitatively read the information of the road in the process of path planning, firstly, the continuous path information needs to be converted into a discrete digital road network that can be easily expressed. In this way, we are able to make good use of these discrete road expressions for path selection and planning. In addition, discretizing the continuous road information can also reduce the computational load and increase the speed of planning. The current search space expression methods include the driving corridor method, the Venn diagram method, the Lattices method and the occupancy grid map method. The driving corridor method consists of a continuous stretch of barrier-free road. The boundary for this road section includes the road and lane boundary, the obstacle points, etc. These road and lane boundary information can be obtained by using the digital map or the SLAM method. The center line in the driving corridor method is the planning path. The Venn diagram method can create a path of the maximum distance between the vehicle and the obstacle point. This algorithm is suitable for use in static environments and can be used in the process of automatic parking. The Venn diagram method often combines the artificial potential field method to generate obstacle avoidance paths. The Lattices method overcomes the efficiency of grid-based algorithms without additional computational load. The occupancy grid map method divides the space into small grids, and then associates each grid with its probability of being occupied by the obstacles or the rate that the grid is used up. Subsequently, performs the path planning according to the probability map or the consumption map. The Lattices method is a continuous process of stacking paths between different states (Fig. 8.1).

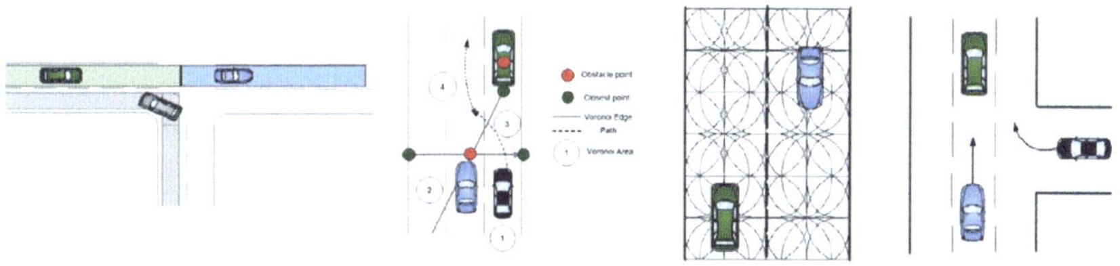

Driving corridor method　Venn diagram method　Lattices method　Occupancy grid map method

**Fig. 8.1** Search space expression method

## 8.3 Path Planning Algorithm

The path planning algorithm of the unmanned system can be divided into the global and the local path planning algorithm. The global path planning algorithm refers to the path planning performed when the system grasps the global environment information. The accuracy of the environmental information obtained by the unmanned systems has a significant impact on the effectiveness of the path planning. The global path planning is able to achieve the optimal solution, but its limitations include intensive computation and relatively poor real-time performance.

The global path planning methods include the configuration space method, the free space method and the grid method. The configuration space method refers to the unmanned systems as a shapeless point. According to the system's length and width, the obstacle undergoes a swelling process, and subsequently path planning is performed such as the use of visual method, the Dijkstra method, and then A* search method. The free space method refers to the representation of free space as a connected graph, and the path planning is performed by means of graph search, which usually has a large time complexity. The grid method refers to the cutting of the environment into equal-sized networks. If there are obstacles in the network, the system cannot pass through the grid.

The local path planning algorithm refers to the path planning of the unmanned system when the environment is unknown or partially unknown. The local path planning algorithm is able to better avoid obstacles and has a better real-time performance and enhanced practicality. However, since the system cannot grasp the global information, it is easy to fall into the local extreme point and exists the scenario when the target position cannot be reached.

Local path planning algorithms mainly include the artificial potential field method, the particle swarm optimization algorithm, the fuzzy logic algorithm, the neural network algorithm and the hybrid method. In the following section, the classic and widely adopted path planning algorithms will be described.

### 8.3.1 A* Search Method

The A* algorithm is one of the heuristic search algorithms. The heuristic search algorithm means that when the current search node is selected, the next node is selected by a heuristic function until the target point is reached. The advantage of the A* algorithm is that a large number of invalid searches can be avoided, thereby improving efficiency. The key to the A* algorithm lies in the design of the evaluation function. The evaluation function is given as follows:

$$f(n) = g(n) + h(n) \tag{8.1}$$

where f(n) is the overall estimate of the node; g(n) is the cost of moving from the starting node to the current node; and h(n) is the cost of moving from the current point to the target node. The following conditions are necessary for the Eq. (8.1) to become the A* search method: (1) there is an optimal path from the starting point to the target point; (2) the problem domain is finite; (3) the cost of all the nodes and sub-nodes is greater than 0; and (4) h(n) is less than the cost of the actual problem. The search algorithm that satisfies the above four conditions is called the A* algorithm, and the optimal solution will definitely be found.

An example of the design of g(n) and h(n) is illustrated below:

Firstly, the green point is the starting point A, the red point is the target point B, the blue point is an obstacle that cannot pass, and the black point is the free area. The goal is to plan the path from A to B. In each square, the top-left corner represents f(n), the lower left corner represents g(n), and the lower right corner represents h(n) as shown in Fig. 8.2c. Wherein, g(n) represents the distance from the parent node (the green starting point in Fig. 8.2a) to the current node; and h(n) is set to the Manhattan distance of the current node to the target node. Obviously, the values of f(n) = g(n) + h(n), g(n), h(n) can be determined by the designer. From the computation, the node with the smallest f(n) is selected which is taken as the current node. After continuous selection, it finally reaches the target point B. The curve of the connected nodes is the optimal selection of the A* algorithm path.

As the existing research further advances, improved versions of the A* algorithms have gradually emerged, such as the dynamic A*, A* that has a specific cost function and the hybrid A*, with the goal of addressing speed and memory related cost problems.

The vehicle state in the 4D discrete grid is represented by the hybrid A*, where two dimensions represent the x–y position of the vehicle center in the smooth map coordinates; the third dimension is the vehicle heading direction θ; and the fourth dimension is the forward or backward motion direction. The hybrid A* search algorithm is guided by two heuristics, also known as the incomplete barrier-free heuristic and the complete heuristic with obstacles. The first heuristic algorithm ignores the obstacles but takes into account the incompleteness of the car. This heuristic can fully pre-calculate the entire 4D space (vehicle position, direction and direction of motion), which helps to approach the target through the required heading in the final plan. The second heuristic algorithm is a twofold method.

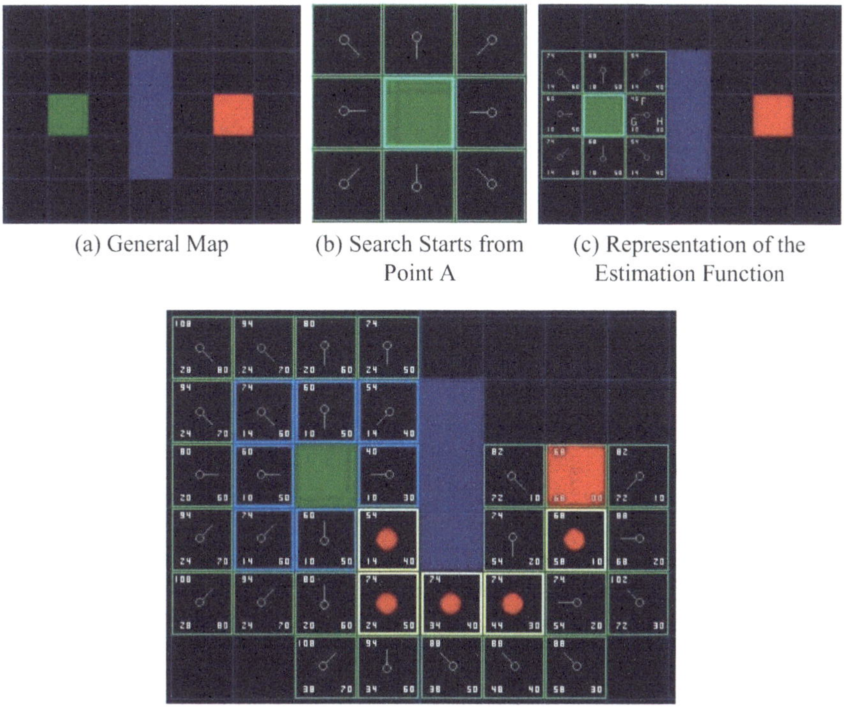

(a) General Map (b) Search Starts from Point A (c) Representation of the Estimation Function

(d) Optimal Path

**Fig. 8.2** A* algorithm implementation process

Firstly, it ignores the incompleteness of the car, but yet calculates the shortest distance to the target. It performs online calculations by performing dynamic programming in 2D (ignoring vehicle direction and direction of motion). Both heuristics are acceptable, so the maximum of the two options will be chosen. Figure 8.3a shows the A* that adopted the commonly used Euclidean distance heuristic. As shown in Fig. 8.3b, the incomplete barrier-free heuristic is significantly more effective than the Euclidean distance heuristic because it takes into account of the vehicle direction. However, as shown in Fig. 8.3c, this heuristic alone fails in the case of a U-shaped dead angle. The plan obtained is very efficient by adding a complete heuristic algorithm with obstacles, as shown in Fig. 8.3d.

### *8.3.2 Artificial Potential Field Method*

Artificial potential field path planning is a virtual force method proposed by Khatib. Its basic idea is to design the movement of the robot in the surrounding environment into an abstract motion in the artificial gravitational field. The target point

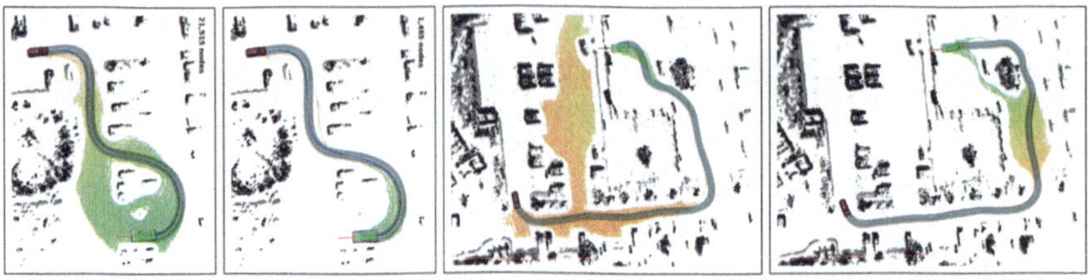

(a) (b) (c) (d)

**Fig. 8.3** Hybrid-state A* heuristics

generates a "gravitational force" to the mobile robot while the obstacle generates a "repulsive force" to the mobile robot to control the movement of the mobile robot. The use of the programming potential field method to produce the path is relatively smooth and is generally safe, however, it exists the problem of the local optima problem.

As illustrated in Fig. 8.4a, the robot is in motion in a two-dimensional environment which indicates the relative position between the robot, the obstacle and the target. In Fig. 8.4b, the constructed virtual artificial potential field is denoted with the initial point of the object located on a higher 'mountain' and the target point to be reached at the foot of the 'mountain'. This forms a potential field where objects that are under the guidance of the potential field could avoid the obstacles and reach the target. The artificial potential field includes a gravitational field and a repulsive field, wherein the target point generates gravity to the object and guides the object toward it. The obstacle creates a repulsive force on the object to prevent the object from colliding with it. The force that the object experiences at each point along the path is equivalent to the total repulsion and gravity forces at each point. Therefore, the key to the artificial potential field method is on how to construct the

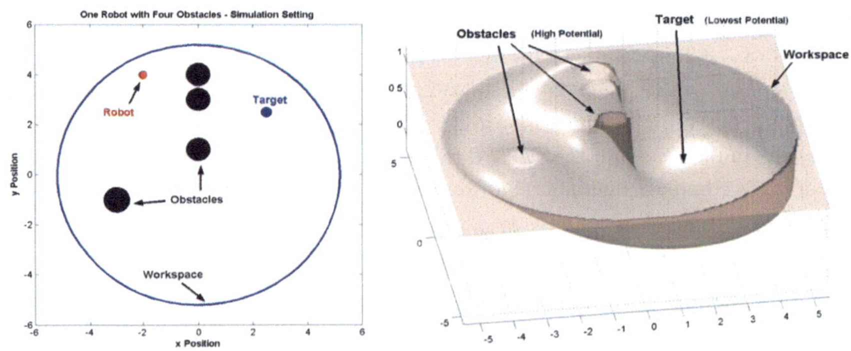


(a) Robot Position (b) Virtual Potential

**Fig. 8.4** Schematic diagram of the artificial potential field method

gravitational field and the repulsive field. For the construction of the gravitational field, the commonly used gravitational functions are:

$$U_{att}(q) = \frac{1}{2}\varepsilon\rho^2\left(q, q_{goal}\right) \tag{8.2}$$

Among them, $\varepsilon$ is the scale factor of $\rho^2\left(q, q_{goal}\right)$ which indicates the distance between the current state of the object and the target. Gravitation is the derivative of the gravitational field against distance (Fig. 8.5):

$$F_{att}(q) = -\nabla U_{att}(q) = \varepsilon\left(q_{goal} - q\right) \tag{8.3}$$

The repulsive field is expressed as:

$$U_{rep}(q) = \frac{1}{2}\eta\left(\frac{1}{\rho(q, q_{obs})} - \frac{1}{\rho_0}\right)^2, \quad \rho\left(q, q_{goal}\right) \le \rho_0 \tag{8.4}$$

$$U_{rep}(q) = 0, \rho(q, q_{obs}) > \rho_0 \tag{8.5}$$

Wherein, $\eta$ is the repulsion scale factor, which $\rho\left(q, q_{goal}\right)$ represents the distance between the object and the obstacle; and $\rho_0$ represents the radius of influence of each obstacle. In other words, by keeping a certain distance, the obstacle has no repulsion effect on the object. The repulsion is the gradient of the repulsive field (Fig. 8.6):

$$F_{rep}(q) = -\nabla U_{rep}(q) = \eta\left(\frac{1}{\rho(q, q_{obs})} - \frac{1}{\rho_0}\right)\frac{1}{\rho^2(q, q_{obs})}\nabla\rho\left(q, q_{goal}\right), \tag{8.6}$$

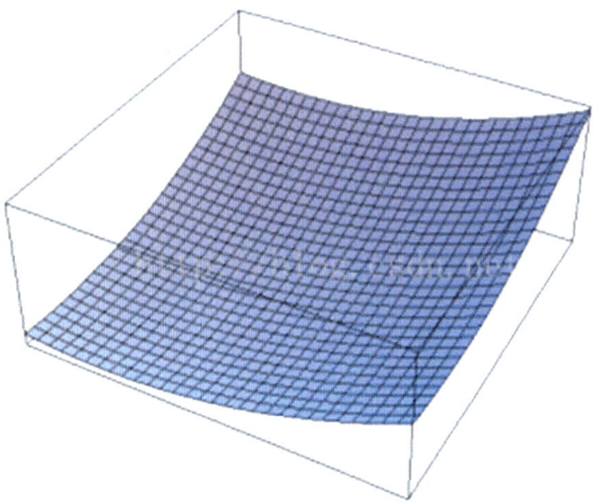

**Fig. 8.5** Gravitational field

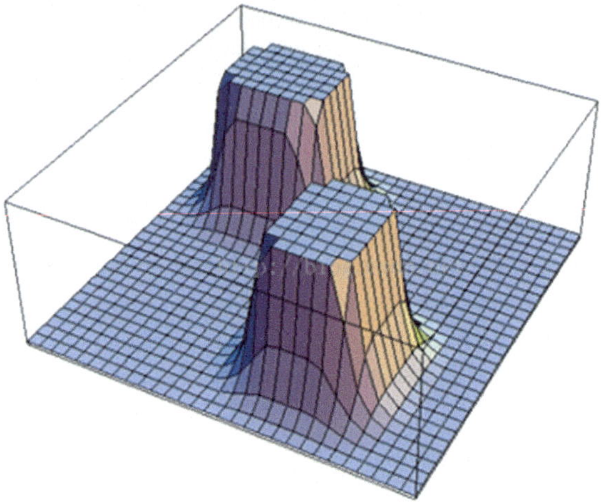

**Fig. 8.6** Repulsive field

$$F_{rep}(q) = 0, \rho(q, q_{obs}) > \rho_0 \tag{8.7}$$

The total artificial potential field is the superposition of the gravitational field and the repulsive field, that is $U = U_{att}(q) + U_{rep}(q)$. The total force is the superposition of the corresponding component forces, as shown in Fig. 8.7:

However, there still exist some problems in this basic potential field. Firstly, when the object is far away from the target point, the gravitation will become extremely large. If a relatively small or negligible repulsive force is present, the object may encounter obstacle in its path. Secondly, when there is an obstacle near the target, there will be a very large repulsive force and the gravity will be relatively small. This makes it difficult to reach the target point of the object. Finally, at some point, when the attraction and repulsion are of equal amount, the direction reverts and the object is prone to fall into a local optimal solution or oscillation.

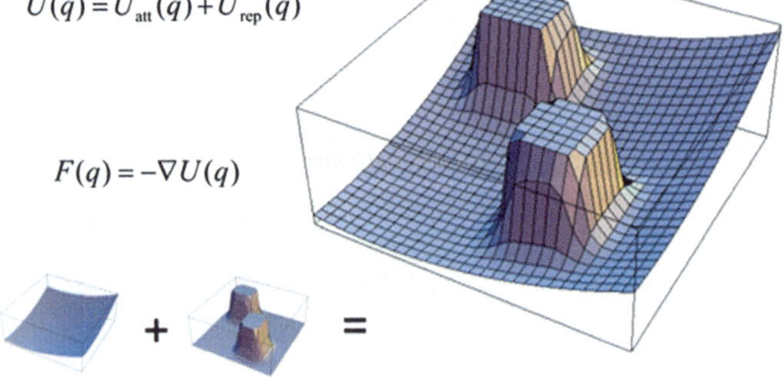


**Fig. 8.7** Artificial potential field

A number of improvements have been proposed for these problems. For problems that may encounter obstacles, the gravitational function can be solved to avoid excessive gravitation due to being too far from the target point.

$$U_{att}(q) = \frac{1}{2}\varepsilon\rho^2\left(q, q_{goal}\right), \quad \rho\left(q, q_{goal}\right) \leq d^*_{goal} \tag{8.8}$$

$$U_{att}(q) = d^*_{goal}\varepsilon\rho\left(q, q_{goal}\right) - \frac{1}{2}\varepsilon\left(d^*_{goal}\right)^2, \quad \rho\left(q, q_{goal}\right) > d^*_{goal} \tag{8.9}$$

In comparison with (8.2), (8.8), and (8.9) increases the range defined. $d^*_{goal}$ sets the threshold which limits the distance between the target and the object. Hence, gravity correspondingly becomes:

$$F_{att}(q) = \varepsilon\left(q_{goal} - q\right), \quad \rho\left(q, q_{goal}\right) \leq d^*_{goal} \tag{8.10}$$

$$F_{att}(q) = \frac{d^*_{goal}\varepsilon\rho\left(q_{goal} - q\right)}{\rho\left(q - q_{goal}\right)}, \quad \rho\left(q, q_{goal}\right) > d^*_{goal} \tag{8.11}$$

Addressing the problem of the inability to reach the target due to the presence of obstacles near the target point, a new kind of repulsion function is introduced:

$$U_{rep}(q) = \frac{1}{2}\eta\left(\frac{1}{\rho(q, q_{obs})} - \frac{1}{\rho_0}\right)^2 \cdot \rho^n\left(q, q_{goal}\right), \rho\left(q, q_{goal}\right) \leq \rho_0 \tag{8.12}$$

Based on the original repulsive field and the influence of the distance between the target and the object, the local optima is the biggest problem in the artificial potential field method. By adding a random disturbance, the object would be able to get out of the local optimal value.

### *8.3.3 Lattices Planning Algorithm*

Under normal circumstances, we use the Cartesian coordinates to describe the position of the object. However, the Cartesian coordinate system cannot fully meet the needs of the unmanned vehicle movement, because even with the given coordinates of the vehicle (x, y), it is still unable to determine the vehicle travel mileage and the distance from the center of the lane. The cartesian coordinate system of alternative solutions is known as the Frenet coordinate system. The Frenet coordinate system describes the position of the car relative to the road. In the Frenet framework, S represents the distance along the road or the vertical axis, and d represents the displacement from the longitudinal line, i.e. the horizontal axis.

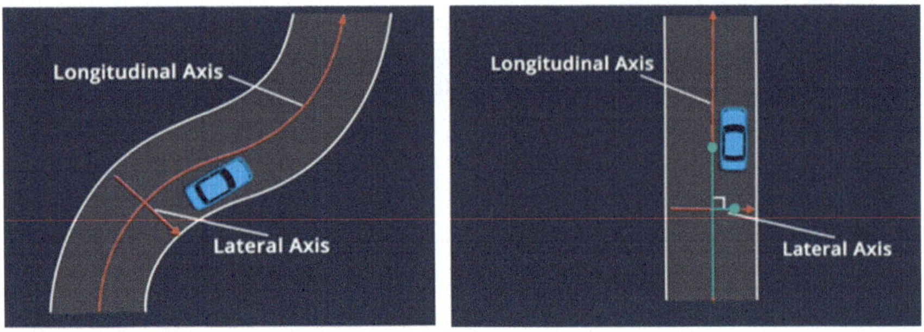


**Fig. 8.8** Frenet coordinate system

At each point of the road, the horizontal and vertical axes are perpendicular to each other. The ordinate indicates the distance travelled along the road, and the abscissa indicates the distance that the car deviates from the center line of the road (Fig. 8.8).

By using the Frenet coordinate system, the environment can be mapped onto the vertical and horizontal axes. If real-time path planning for unmanned vehicles is to be performed, the three-dimensional trajectories, that is the vertical dimension, horizontal dimension and time dimension, need to be generated. The three-dimensional problem can be decomposed into two separate two-dimensional problems by separating the longitudinal and lateral components of the trajectory. One of the two-dimensional trajectories is a time-longed longitudinal trajectory called ST trajectory, and the other two-dimensional trajectory is relative to the longitudinal trajectory. The lateral offset is called the SL trajectory (Fig. 8.9).

Lattice planning involves two steps which creates the ST and SL trajectories separately before combining. To generate the longitudinal and lateral two-dimensional trajectories, the initial state of the vehicle needs to be transmitted into the ST coordinate system and the SL coordinate system, and the final vehicle state is selected by sampling the candidate final states in the pre-selected mode. A set of trajectories is constructed for each candidate final state. The vehicle is then converted from its initial state to its final state, and the trajectories are evaluated using a cost function. The lowest cost trajectory is selected as the last selected optimal path.

**Fig. 8.9** ST and SL trajectories

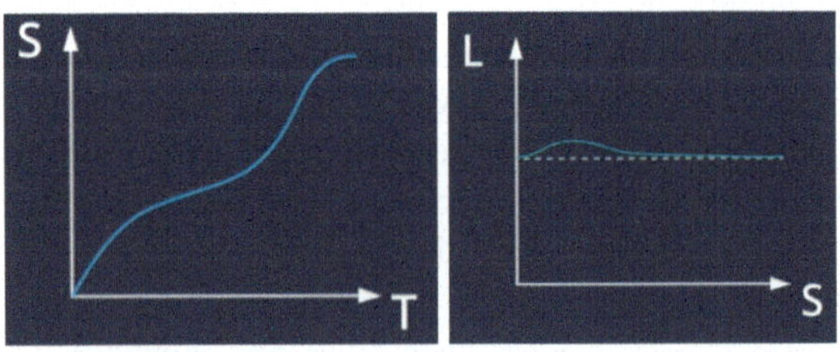

### 8.3.4 RRT Path Planning Algorithm

The Rapidly Exploring Random Trees (RRT) algorithm is an efficient randomized path planning algorithm for solving high dimensional spatial path planning problems. The advantage is that it can directly be used to solve the path planning problem of non-integrity constraints. By randomly sampling the state space, the complex modelling of the environment space is avoided, which is suitable for solving the path planning of multi-degree-of-freedom robot under complex constrained problem.

The basic idea of the algorithm is to randomly generate a state sample, and subsequently perform the forward prediction of the control system. The trajectory is predicted with the obstacle map for safety testing. The feasibility of the trajectory would then decide if the sampling point could be used. The RRT attempts to expand to a randomly selected new state at each step. Firstly, the starting node, $x_{init}$, is selected as the root node of the tree in the task area. The random tree is constructed by expanding the leaf nodes from the root node. Randomly selected state, $x_{rand}$, is selected in the free space region. A leaf node that is nearest to $x_{rand}$ among all the current leaf nodes in the random tree is selected, and is known as the neighbour node, $x_{near}$. Thereafter, a step distance, $\varepsilon$, is extended from $x_{near}$ towards $x_{rand}$ to obtain a new node, $x_{new}$. During the extension process, it is determined whether there is a conflict with the known threat area. If there is no conflict, the new node, $x_{new}$, is accepted and added as a node in the random tree. If there exists conflict between the new node, $x_{new}$, and the threat area, this means that the new node from the extension does not meet the safety requirement. Hence, the new node is discarded and a new random node, $x_{rand}$, is selected. Through the continuous iterated extension, when the leaf node of the random tree is close enough to the target position, the construction of the random tree is considered to have completed. At this time, the leaf node that is closest to the target position is taken as the starting position, and the parent node is searched upwards. A feasible path from the starting position to the target position in turn is obtained. The size of the step has a certain impact on the planning efficiency. In a simple threat environment, a larger step size can be used to speed up the planning, and the resulting path is shorter. In a complex threat environment, an excessive step size limits the scalability of the RRT algorithm, resulting in a lower planning success rate. Moreover, when the length of the obtained path is large, it does not meet the optimality requirement. At this juncture, a smaller step size can be selected (Fig. 8.10).

In the RRT construction process, on one hand, it is hoped that the growth of the random tree can have a tendency towards the target position so as to promote the growth of the random tree toward the target position (vertical). In doing so, the rapid growth of the random tree is facilitated towards the vicinity of the target position. On the other hand, the random trees need to have the ability to evade threat areas, which in turn requires the random tree to scale horizontally within the task environment, with the strong spatial exploration capabilities. In the RRT algorithm, the method of choosing the random points, $x_{rand}$, and step sizes affect the

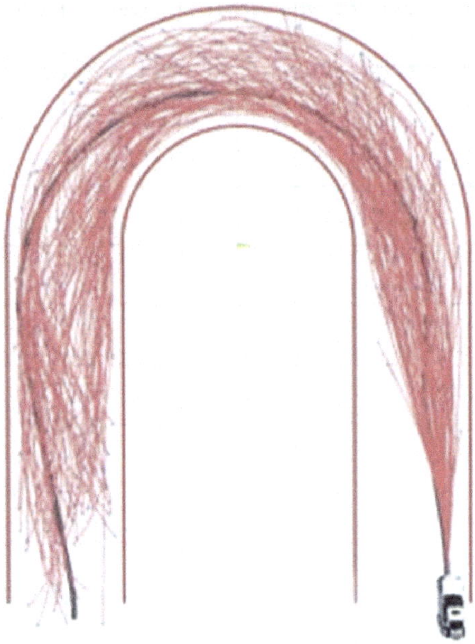

**Fig. 8.10** RRT path planning

quality of the planning. Chaotic variables can be used to generate the location of the random nodes, $x_{rand}$, within the task environment of the RRT path planning. Chaos is the general term for the random behavior exhibited by deterministic systems. The root cause is the nonlinear interaction within the system. Chaotic systems are characterized by its boundedness, ergodicity and randomness. The motion state generated by the chaotic system under external deterministic excitation is random. The sensitivity of the chaotic system to the initial value is reflected in the local instability of the system, which is also the cause of this inherent randomness. Due to these special properties of chaotic systems, the chaotic state sequence generated by the chaotic system can be used as the initial value of the optimization algorithm when solving the optimization problem. The use of chaotic sequences to generate random nodes can guarantee the randomness of the random node selection, and can fully utilize the ergodicity and regularity of the chaotic dynamics. Random nodes in the process of random tree growth cover the whole task environment as much as possible. For the step size, the fuzzy inference system can be used to dynamically set the step size. The fuzzy inference process is a simulation of the human reasoning process based on the fuzzy concepts by the computer. The number of fuzzy partitions determines the maximum number of the fuzzy rules. If the number of the partitions is too small, the fuzzy reasoning will be too rough. As the number of segments becomes larger and more refined, the accuracy of reasoning becomes higher as well, but it will also lead to a surge in computation. At present, the number of fuzzy segments is generally obtained based on experience.

### 8.3.5 Genetic Algorithm

Genetic algorithms operate on a limited set of parameter length coding, and is adapted to generate a value for each of the encoded code to evaluate its quality. By encoding and adapting the values, the algorithm does not need to consider the specifics of the problem. However, for a specific problem that is difficult to encode, it is not necessary to do so. Alternatively, it is entirely possible to directly operate on the parameter set of the problem instead. The basic genetic algorithm is divided into two crossover recombination and replication steps. Copying is conducted according to the quality probability of each parameter set in a class of parameter sets. The copied parameter set is then prepared for cross reorganization. The crossover recombination facilitates the interchanging of a particular portion in each of the two parameter sets to generate a new set of parameters. After copying and crossover recombination, by survival of the fittest mechanism, the badly performed parameters are eliminated and thus the performance of the parameter set is change. With the new set of parameters, a new round of replication and crossover recombination continues. In order to speed up the search, the genetic algorithm also introduces concepts such as "mutation" and "migration". Mutation means to artificially mutate the parameters in a parameter set to achieve the purpose of generating new variants. Migration means to add new parameter sets artificially. Both mutation and migration are conducted with a small probability.

When the genetic algorithm solves the combinatorial optimization problem, the feasible solution of the problem forms the chromosome, randomly selects some individuals to generate the initial population, and then transforms the objective function of the combinatorial optimization problem into the fitness function. The individuals are selected based on the calculated value of the fitness function. Finally, the individuals are selected or cross-mutated to produce individuals with higher fitness values. By the continuous progeny, the offspring are more adapted to the environment until the desired termination requirement is satisfied, thereby, forming an optimal solution for the population. The steps in solving the optimization problem with the genetic algorithm are fixed, summarizing the population initialization, selection, crossover, mutation, and fitness calculation.

Genetic algorithms have a strong global search ability, especially when the crossover probability is relatively large, which can generate a large number of new individuals and improve the global search range. However, the traditional genetic algorithm solves the problem of path planning for mobile robots, and there is a premature convergence problem, and the effect of path planning is unstable. Therefore, at this stage, the combination of other algorithms is used to make up for the shortcomings of the genetic algorithm (Fig. 8.11).

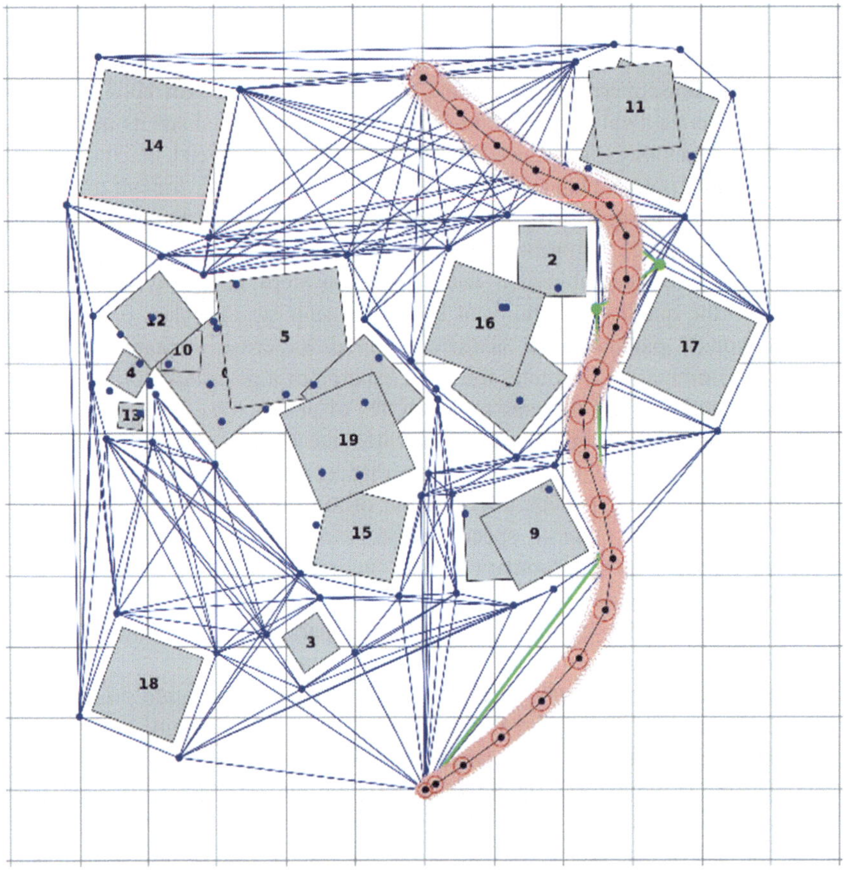


**Fig. 8.11** Genetic algorithm path planning

## 8.3.6 Ant Colony Algorithm

The ant colony algorithm is obtained by simulating ant foraging activities in real life. In real life, based on the pheromone that the ant released on the path, after a period of time, the entire ant colony will find the closest path between the cave and the food through these pheromones. When an ant discovers a pheromone that the ant left on the path, it will particularly prefer a path with a high concentration of these pheromone. If there are two paths of different lengths between the cave and the food, there is no pheromone between the cave and the food at the time of initialization, and any ant releases the same pheromone. The ants that have just started have no inspirational information. Each ant leaves a pheromone along the way. The number of ants choosing different paths is the same. However, the ants that pass the shorter path return to the nest first compared to the ants that have

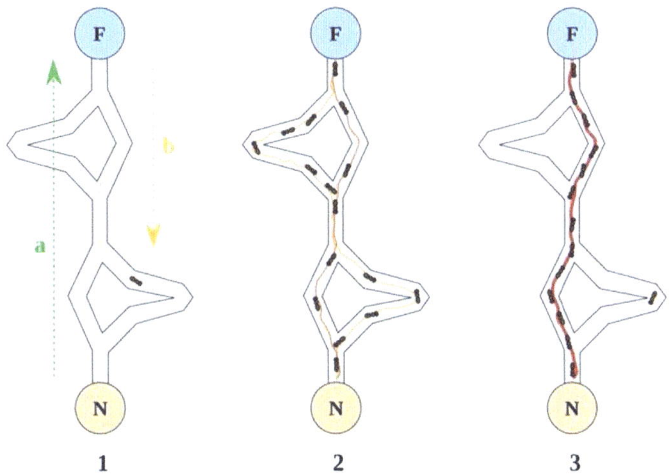


**Fig. 8.12** Ant colony algorithm similar to ant pathfinding

passed the longer path. After a period of time, the shorter path passes more ants and leaves more pheromones, which is inspired by information for the ants that start later, preferring to choose shorter paths. Based on this positive feedback mechanism, in the end, all ants can choose a shorter path (Fig. 8.12).

Combined with the idea of ant foraging, the most commonly used ant colony algorithm is based on the raster map. The distance between the raster maps is in Euclidean distance:

$$d_{i,j} = \sqrt{(x_i - x_j)^2 + (y_i - y_j)^2} \tag{8.13}$$

wherein, $(x_i, y_i)$ and $(x_j, y_j)$ represent the coordinate values of the grid i and the grid j in the grid map, respectively.

$$W = \sum_{i=0}^{M} \sum_{j=0}^{N} b_{i,j}(t) \tag{8.14}$$

$b_{i,j}(t)$ indicates the total number of ants at the coordinate point $(x_i, y_i)$ at time t.

During the movement, the ants $k(k = 1, 2 \ldots W)$ will determine the direction of the next move according to the amount of pheromone stored on the surrounding grid and the corresponding distance. At time t, the transition probability of ant k that needs to move from the position point ($x_i$, $y_i$) to the next position point $(x_j, y_j)$ is given as:

$$p_{i,j}^{k}(t) = \begin{cases} \frac{\tau_{i,j}^{\alpha}(t)\eta_{i,j}^{\beta}}{\sum_{r \in s_j^k}\left(\tau_{i,j}^{\alpha}(t)\eta_{i,j}^{\beta}\right)}, & j \in s_i^k \\ 0 \end{cases} \tag{8.15}$$

$\tau_{i,j}(t)$ represents the amount of pheromone that is left on the path at time t from grid $(x_i, y_i)$ to grid $(x_j, y_j)$. At initialization, the amount of information of the adjacent interval on every path is set at an equal constant C. $\eta_{i,j}^{\beta}$ represents the partial heuristic function for visibility (also defined as $1/d_{i,j}$); parameters α and β represent the weight of influence on the transition probabilities, $\tau_{i,j}^{\alpha}(t)$ and $\eta_{i,j}^{\beta}$, respectively; and $s_k^i$ represents the feasible directions of the *K*th ant at position $(x_i, y_i)$. As time passes, the pheromone left by the ants on the path will gradually evaporate. The legacy formula of pheromone is:

$$\tau_{i,j}(t+1) = (1-\rho)\tau_{i,j}(t) + \Delta\tau_{i,j}(t, t+1) \tag{8.16}$$

$$\tau_{i,j}(t, t+1) = \sum_{k=1}^{m} \Delta\tau_{i,j}(t, t+1) \tag{8.17}$$

Wherein, $\tau_{i,j}(t+1)$ represents the amount of pheromone that is left on the path at time $t+1$ from grid $(x_i, y_i)$ to $(x_j, y_j)$ along the path at time t + 1; $p \in [0, 1]$ represents the pheromone volatilization coefficient; $\Delta\tau_{i,j}^{k}(t, t+1)$ represents the pheromone quantity for the kth ant that stays on the path from the grid $(x_i, y_i)$ to $(x_j, y_j)$ from time t to t + 1; and $\Delta\tau_{i,j}(t, t+1)$ represents the sum of the pheromone received through the raster on the path that passes from the grid $(x_i, y_i)$ to $(x_j, y_j)$ at time t + 1. The volatilization of pheromones can avoid the problem of path sub-optimization caused by too much pheromone on the path.

## 8.4 Application of Path Planning in Unmanned Systems

### *8.4.1 Application in Unmanned Aerial Vehicle*

An important application in the unmanned aerial vehicle comprises the reconnaissance and inspection functions. For instance, the transmission lines are distributed across the regions with many points, complex terrain in a harsh natural environment. The transmission lines and equipment undergo long-term exposure in the field which are subjected to mechanical tension, lightning strikes, material aging, ice coating and other human factors that resulted in phenomena such as inverted towers, broken strands, abrasion, corrosion and movement. These conditions must be repaired or replaced in time. Modern drones have the ability to work at high altitude, over long distance, is fast and self-operating. By planning a

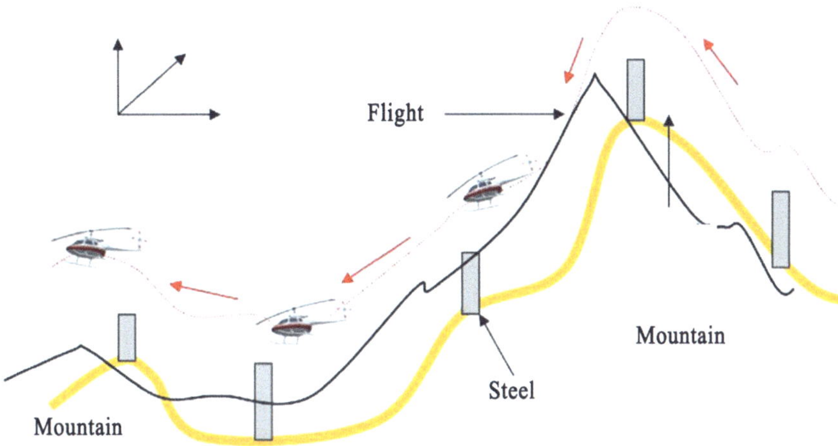


**Fig. 8.13** Schematic diagram of drone inspection

reasonable flight route, drones can cross the mountains and rivers to quickly survey the transmission lines, towers, brackets and the overhead lines to perform a full-spectrum of fast imaging and fault monitoring. Based on the professional analysis of the transmission line surveillance data collected by the UAV, it provides data support for the power line management and maintenance.

UAV operations can greatly increase the speed and efficiency of the transmission maintenance and overhaul, enabling many jobs to be completed quickly in a fully charged environment. UAV operations also allow for a rapid expansion of work and are not plagued by sludge and snow. Therefore, the drone line inspection method is undoubtedly a safe, fast, efficient, and promising line of inspection. Figure 8.13 shows the schematic diagram of the drone inspection, in which the red dotted line is the planned flight path of the drone.

## 8.4.2 Application on Unmanned Vehicles

The main application of the global path planning in the unmanned vehicles is to achieve automatic driving within a pre-set range. In the unmanned vehicle, the global path planning typically relies on the topological maps. Depending on the start and the end points of the vehicle, the control system records the requirements and generates a vehicle travel path. The unmanned vehicle then drives based on such an electronic sensing path. In this process, when the vehicle encounters new obstacles, it will automatically select the brakes, change the speed and the direction according to the local path planning to realize the function of the automatic driving. According to the requirements of the scene, the vehicle can select a suitable motion track. Figure 8.14 illustrates the global path planning in an unmanned vehicle.

(a) (b)

**Fig. 8.14** **a** Schematic of the global route planning in an unmanned vehicle, **b** schematic of partial vehicle route planning in an unmanned vehicle

The three DARPA Driverless Challenges held by the US Defense Advanced Research Projects Agency (DARPA) in 2004 are important milestones in the development of modern unmanned driving. Many path planning algorithms have been validated in the DARPA Series Challenge. For example, in the DARPA Grand Challenge in 2005, Alice's unmanned vehicle motion planner used the optimal time as the index, and adopted the vehicle speed, acceleration, steering wheel angle, steering speed and rollover as constraints to realize real-time dynamic obstacle avoidance trajectory planning. In 2007, in the DARPA Urban Challenge, University of Pennsylvania's unmanned vehicle, Little Ben, uses the Dijkstra algorithm. MIT's Talos unmanned vehicle that is based on RRT motion planning methods had successfully implemented the lane keeping, K-Tum, U-Tum, change lanes for overtaking, and congestion avoidance. Stanford University's unmanned vehicle, Junior, had based on the needs of the game environment and used a hybrid A* algorithm which achieved runner-up of the game. Carnegie Mellon University developed the unmanned vehicle, Boss, which adopted the AD* algorithm and eventually won the championship. After the DARPA Series Challenge, path planning algorithms are becoming more widely used in unmanned vehicles. The laboratory of LIVIC (The Laboratory on Interactions between Vehicles, Infrastructure and Drivers) used the polynomial curve for the lane changing scene of the unmanned vehicle, that is, the fifth-order polynomial to describe the lateral motion, and the fourth-order polynomial to describe the longitudinal motion. The Toyota North American Research Center establishes a nonlinear optimization model based on the path node of the hybrid A* plan. The distance, curvature and path smoothness between the path node and the obstacle are used as the optimization indicators. The conjugate gradient algorithm is used to solve the optimization problem and obtain a smoother path. In 2013, Mercedes-Benz S-class intelligent car, BERTHA, completed 100 km of automatic driving, using the numerical optimization algorithm to plan the motion trajectory. The algorithm used the relative position, acceleration, yaw rate and curvature of the vehicle in the lane as the optimization index. The trajectory calculation is performed by the sequential quadratic programming method. In 2016, Baidu's open source apollo autopilot

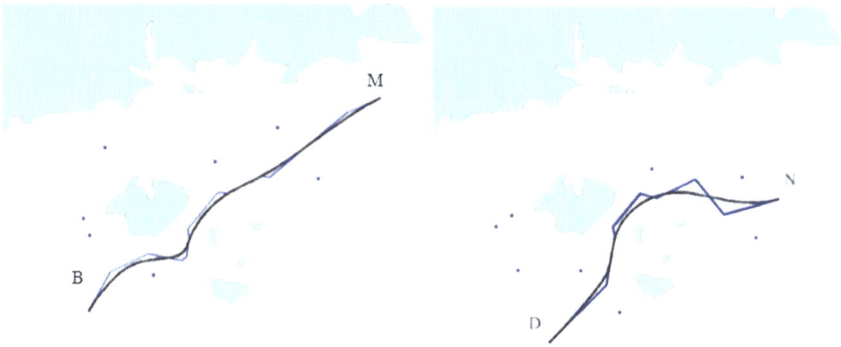


**Fig. 8.15** Path planning in unmanned surface vehicle

platform uses the maximum expectation algorithm (EM) and the lattice algorithm based on curve interpolation to realize real-time planning of the unmanned vehicles on complex or simple urban roads.

### *8.4.3 Application in Unmanned Surface Vehicle*

The unmanned surface vehicle has a wide range of application. Some of the main application include water surface surveys, seabed search and exploration, navigation assistance, vessel tracking and other tracking missions. For all missions, path planning is an important step of ensuring that the unmanned surface vehicle can sail safely to successfully complete the missions. Under the overall path planning, the unmanned surface vehicle looks for a navigation path that meets certain evaluation criteria from the known starting point to the target in the working environment where static and dynamic obstacles coexist, so that the unmanned surface vehicle is able to safely and reliably avoid the obstacles during the navigation process. Through the optimal path planning problem, the navigation reliability, efficiency and energy consumption of the unmanned boat are realized (Fig. 8.15).

## Bibliography

1. González D, Pérez J, Milanés V et al (2016) A review of motion planning techniques for automated vehicles. IEEE Trans Intell Transp Syst 17(4):1135–1145
2. Werling M, Ziegler J, Kammel S et al (2010) Optimal trajectory generation for dynamic street scenarios in a Frenét frame. In: IEEE international conference on robotics and automation. IEEE, pp 987–993
3. Chai Y, Tang QH, Deng MX et al (2016) Raster model construct of the robot path planning with ant colony algorithm. Mech Des Manuf 4:178–181

4. Zhang H (2013) Research on path planning algorithm of ground autonomous mobile robot [D]. Zhejiang University, Zhejiang
5. Zhang G, Hu X, Chai J, Zhao L, Yu T (2011) Overview of machine application of path planning algorithm. Mod Mach 5:85–90
6. Xu P (2009) Research on global path planning algorithm for mobile robots in complex dynamic environment [D]. School of Automation, Beijing University of Posts and Telecommunications, Beijing
7. Wang T (2009) Research on robot path planning and simulation system based on ant colony algorithm [D]. Xi'an University of Science and Technology, Xi'an
8. Khatib Oussama (1986) Real -time obstacle avoidance for manipulators and mobile robots. Int J Robot Res 5:90–98
9. Siciliano B, Sciavicco L (1998) A solution algorithm to the inverse kinematic problem for redundant manipulators. IEEE J Robot Autom 4:403–410
10. Tan K, Sun M, Sun C (2003) Robot motion planning based on improved artificial potential field in dynamic environment. J Shenyang Univ Technol 5:568–570
11. Du Z, Liu G (2009) Mobile robot path planning based on genetic simulated annealing algorithm. Comput Simul 26(2):118–121
12. Romeijn HE, Smith RL (1994) Simulated annealing for constrained global optimization. J Global Optim 5(2):101–124
13. Chen LN, Aihara K (1995) Chaotic simulated annealing by a neural network model with transient chaos. Neural Networks 8(6):915–930
14. Shang DY (2010) A fuzzy window path planning method based on fuzzy logic. Numerical Technology 2:146–148
15. Memon KR, Memon S, Memon B et al (2016) Real time implementation of path planning algorithm with obstacle avoidance for autonomous vehicle. In: International conference on computing for sustainable global development. IEEE
16. Li J, Deng G, Luo C et al (2016) A hybrid path planning method in unmanned air/ground vehicle (UAV/UGV) cooperative systems. IEEE Trans Veh Technol 65(12):9585–9596
17. Chen Zhijun, Zeng Z (2018) Three-dimensional path planning of robot based on fuzzy neural network and genetic algorithm. J Chongqing Normal Univ (Natural Science Edition) 1:93–99
18. Wu B, Luo F (2017) Intelligent vehicle routing planning algorithm based on RRT. Mechatronics 10:15–23
19. Feng L, Liang H, Du M et al (2017) RRT intelligent vehicle path planning algorithm based on A* guidance domain. J Comput Syst 26(8):127–133
20. Jin H (2016) Research on optimization of vehicle path planning algorithm for urban intelligent transportation system. Inf Syst Eng 12:44

The manufacturer's authorised representative in the EU is Springer Nature Customer Service Centre GmbH, Europaplatz 3, 69115 Heidelberg, Germany. If you have any concerns regarding our products, please contact ProductSafety@springernature.com

Printed and bound by CPI Group (UK) Ltd, Croydon, CR0 4YY
27/07/2026
02177351-0003